A BIRDWATCH
GUIDE TO THE
CANARY ISLANDS

Tony Clarke and
David Collins

Illustrations by Phil Jones

**BIRD
WATCHERS'
GUIDES**

Prion Ltd.
Perry

CONTENTS

ACKNOWLEDGEMENTS

We would like to thank the following persons, all of whom assisted in the preparation of this book. Juan Antonio Lorenzo Gutiérrez assisted with the bird list and commented on the first draft. Peter Crocker, Alison Collins and Marcus Kohler also commented on the first draft, and Marcus Kohler also assisted with a final reconnaissance of Fuerteventura and Lanzarote. Garry Tapper provided information on sites on El Hierro, and Keith Emmerson provided information on sites on La Palma. Lists of mammals and reptiles/amphibians were checked by Aurelio Martín, and the list of whales and dolphins was checked by Roberto Montero Lopez.

Finally, we would like to thank our wives Luisa Santos García and Alison Collins for putting up with us during preparation of the guide.

Tony Clarke

David Collins

INTRODUCTION

The Canary Islands are now firmly established as a 'must' for any birdwatcher with an interest in Western Palearctic birds. Although only 92 species breed or have possibly bred in the island group, and 12 of these are introductions, five are endemic to the islands and a further two species are confined to Macaronesia. Macaronesia includes five groups of islands in the central part of the eastern Atlantic. From north to south these are: the Azores, Madeira, the Salvage Islands, the Canary Islands and the Cape Verde Islands. All of these islands are volcanic in origin, and they show certain botanical and zoological affinities. The Canary Islands is the closest of these island groups to Africa. Fuerteventura, which is the most easterly of the Canary Islands, is only 100km from the coast of southern Morocco. In addition to the endemics, several scarce seabirds breed on the islands and a number of desert species are resident. This combination of birds, together with the spectacular scenery, make a visit to the Canary Islands an unforgettable experience for any birdwatcher.

For botanists the islands are an absolute paradise with more than 2,000 species of plants, of which about 500 are endemic. Of particular note is the range of endemic succulents (especially spurges) and laurel trees. There is also an interesting, if somewhat limited, range of butterflies, moths and dragonflies, and there are a number of endemic lizards (but no snakes). The mammals, with the exception of bats and sea mammals, are all introduced except for an endemic shrew. At the end of this guide there are complete lists of butterflies, reptiles and amphibians, and mammals, as well as an annotated list of bird species.

The Canary Islands can be described as sub-tropical, and at least at lower levels, the climate is mild and frost free in winter, but relatively cool in summer due to the moderating influence of the sea. However, all of the islands are to some extent mountainous, and the highest areas (on Tenerife) are sub-alpine in character with permanent winter snow.

From an ecological point of view the Canary Islands can be divided into two groups. The Eastern Islands (Fuerteventura, Lanzarote and neighbouring islets) are relatively low-lying, with no mountains above about 800m. With the exception of the highest peaks, they are very arid with a more or less uniform cover of semi-desert scrub. They have no proper woodland and are therefore lacking in certain bird species which are characteristic of the other islands. The other islands (Gran Canaria, Tenerife, La Gomera, La Palma and El Hierro) can all be included in the Western Islands, although Gran Canaria is politically part of the Eastern Islands group. All of these have laurel forest and Gran Canaria, Tenerife and La Palma also have forests of the endemic Canary Islands Pine.

Tenerife is the largest of the islands (2,057km²) as well as the highest, while Fuerteventura is the second largest (1,731km²). El Hierro is the smallest of the main islands (277km²), but there are also six islets, all of which form part of the Eastern Islands group. One of these (Lobos) lies off the north coast of Fuerteventura, whereas the other five (La Graciosa, Montana Clara, Alegranza, Roque del Oeste and Roque del Este) lie to the north of Lanzarote.

In geological terms the Canary Islands are quite recent. The Eastern

Islands are judged to be about 16-20 million years old. Moving west the islands become younger, and La Palma and El Hierro are only about 2-3 million years old. There is some evidence that Lanzarote and Fuerteventura were once joined to Africa, although the evidence to the contrary is usually judged to be the stronger. Fossilised eggshells of a large bird have been discovered on Lanzarote. These were originally thought to be the eggshells of Ostriches or Elephant-birds, but it is now thought that they could be from more primitive birds which may have possessed the power of flight. It is also possible that the eggs were brought to the island by man. They do not, therefore, confirm that Lanzarote was connected to the African mainland.

There have been volcanic eruptions on several of the islands in historic times, mainly on Lanzarote, Tenerife and La Palma. The most recent was on La Palma in 1971, while on Tenerife there were eruptions in 1798 and 1909.

It is possible that the Canary Islands were uninhabited before Roman times. The origin of the pre-historic inhabitants (Guanches) is open to speculation, but the available evidence appears to suggest (not surprisingly) that they were related to the Berbers of North Africa. There are interesting Guanche remains in all of the islands (especially Gran Canaria and Fuerteventura) in the form of caves and there are several Guanche museums. However, there are no written records for the years before the discovery of the islands by Europeans, and subsequent conquest by the Spanish in the 15th century. Tenerife was the last island to fall to the invaders, being conquered by Alonso Fernandez de Lugo between 1494 and 1496. Since then the islands have been under Spanish rule. However, pirates (including British ones) took their toll from time to time, and an attempted conquest by Nelson in 1797 was forced away from Santa Cruz de Tenerife (resulting in the loss of Nelson's arm).

There has been much debate about the origin of the name 'Canary' Islands, as well as the individual names of several of the islands. The name 'Isla Canaria' first appeared on a Spanish chart in 1339. Since Roman times the island group was known as the Blessed or Fortunate Islands. Pliny the Elder (AD 23-79) used the name 'Capraria' for Lanzarote, meaning the one which swarms with lizards. 'Nivaria' meaning snow-covered was the name for Tenerife. He used the name 'Canaria' for Gran Canaria, possibly in reference to the dogs (Latin canis = dog) which lived on the island. However, the Romans possibly associated the islands with the kingdom of the dead which, according to their beliefs, was in the west. In Roman mythology the dead were taken into the underworld by dogs. Other suggestions for the origin of the name are that the bird known to the Romans as 'canora' (singing bird) may have occurred on the islands. It is also possible that the name may have been taken from Cape Canauria (probably Cape Bojador) on the African coast.

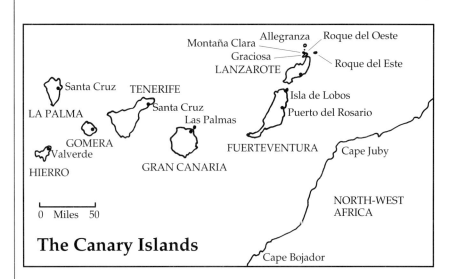

The Canary Islands

PRE-TOUR INFORMATION

Visas and passports

Politically, the Canary Islands are part of Spain, and entry requirements are the same as for that country. In theory, this means that visitors from other European Union countries only need to produce valid identification, although it is advisable to carry a current passport. A passport is essential for non-EU citizens. No visa is required for European or North American visitors staying for less than three months. For longer stays it is best to contact the Spanish Embassy for advice. In the UK, details can also be obtained from the Spanish National Tourist Office, 57-58 St. James Street, London, SW1A 1LD. (Tel: 0171-499-0901). No inoculations are required to gain entry to the country from Europe, North America, Australia or New Zealand.

Currency and Exchange Rate

The currency is the Spanish Peseta. In 1996 the exchange rate was around varied from 182 to 195 pesetas to the pound (sterling). In the main tourist areas there are plenty of places where money can be exchanged, including various shops as well as hotels and banks. In these areas, it is sometimes possible to pay using sterling notes if absolutely necessary. However, in the more remote areas, exchange is only possible at banks, and it is then necessary to be aware of the local bank opening times.

Field Guides

The only English language field guide to cover the area properly is the 'Birds of Britain and Europe with the Middle East and North Africa' by Heinzel, Fitter and Parslow, although a field guide to the eastern Atlantic Islands is currently being prepared. However, there is now a specific guide to the birds of the Canary Islands in Spanish. This is 'Guia de las Aves de las Islas Canarias' by Jose Manuel Moreno. It includes plates of all breeding and migrant species, together with all vagrants recorded up to 1992. Another book in Spanish which is invaluable if visiting Tenerife is 'Atlas de las Aves Nidificantes en la Isla de Tenerife' by Aurelio Martin. These last two books are available through specialist natural history book suppliers, and while they can be bought in bookshops in the larger cities in the islands, it is advisable to purchase them before travelling.

The only ornithological text book specific to the birds of the Canary Islands is 'Birds of the Atlantic Islands Vol.2: The Canary Islands' by David Bannerman. This is a lovely book to read, but it is long out of print and all but unobtainable, except from some public libraries. Although it is now out of date in many respects, it is well worth reading if you can find a copy. The standard ornithological text, as for anywhere else in the region, is the nine volume 'Handbook of the Birds of Europe, the Middle East and North Africa' by Cramp et al.

Emergencies

In case of emergencies, the British Consulate can be contacted (between 08.30 and 13.30) at:

Plaza Weyler No.8, 38003 Santa Cruz, Tenerife. (Tel: 28-68-63).

Luis Morote No.6, 3rd floor, 35777 Las Palmas, Gran Canaria. (Tel: 26-25-08).

TRAVEL INFORMATION

There are airports on all of the Canary Islands except La Gomera where one is currently under construction and due to be finished sometime in 1997. Charter flights from the UK and continental Europe are frequent to Tenerife, Gran Canaria and Lanzarote but there are fewer flights to Fuerteventura. At present, La Palma has charters from Germany only. El Hierro is now the only island which is more or less untouched by tourism. Iberia operate scheduled flights to the Canary Islands from the Spanish mainland, and Monarch operate scheduled flights to Tenerife from Luton Airport in the UK. A car ferry operates from Cadiz (mainland Spain).

If travelling with a holiday company, the travel representative will almost certainly reconfirm your return flight on your behalf. However, if travelling independently, remember that you must reconfirm your return flight at least two days before departure.

Car Hire

By far the easiest way to get around any of the islands is by hiring a car for the duration of the holiday. The driver must be at least 21 years old and in possession of a current driving licence. On Tenerife, Gran Canaria and Lanzarote cars are available from numerous outlets and at varying prices, but the choice is more limited on Fuerteventura. On La Gomera, La Palma and El Hierro cars can only be hired from the airport or harbour areas, and from the few hotels. For these islands (and for Fuerteventura) it is definitely best to book a car in advance, since it often difficult to find one at short notice.

One of the more popular times for birdwatchers to visit the Western Islands is August, which happens to be one of the busiest months for tourism in general. If visiting at this time it is advisable to book a hire car well in advance as there is often a lack of available vehicles. This is probably best done when booking the flight.

The major companies such as Avis, Hertz, Betacar and Cicar are represented at most of the island airports, and have a wide range of vehicles, including mini-buses and four-wheel drives. Car hire is fairly cheap, and by shopping around before departure it is possible to hire a small car for a week (inclusive of insurance and unlimited mileage) for about £135-180 (1996 prices). Remember that not all petrol stations are open on Sundays, most of them close at night, and not all accept credit cards, and none accept American Express.

Buses and hitch-hiking

If car hire is outside your budget, then many of the sites in Tenerife are accessible by bus or by hitch-hiking. However, on the less developed islands, bus routes do not cover all areas and buses tend to be very infrequent. Car hire is certainly invaluable on Fuerteventura where buses are very limited and there is little other traffic, which makes hitch-hiking time consuming.

Sea and air

Travelling between islands can be achieved in one of two ways, by sea or by air. Ferries run between all the islands but the times and days of sailing change fairly regularly, so check these on arrival. The most important ferry for birdwatchers runs between Los Cristianos, Tenerife and San Sebastian de la Gomera, as this is the one used for seawatching. Two ferries operate to Gomera at the moment and the Benchijigua 2 of the Ferry Gomera company departs from Los Cristianos at 09.00, 12.30, 15.30 and 20.00, returning from San Sebastian at 10.45, 14.15, 18.00 and 07.00. In 1996 the passenger fare varied from 3,000-4,200 pesetas return, depending on sailing time. There are also hydrofoils and jetfoils running between many of the islands, but these are too fast for seeing seabirds.

The inter-island air service is very good, with regular flights between all the islands. Flights are only a little more expensive than the ferries. From Tenerife most internal flights leave from the northern airport (Los Rodeos), but occasional ones operate from the southern airport (Reina Sofia). Check the departure and return airports before flying as it is not unknown to leave from Los Rodeos and return to Reina Sofia. Also, inter-island flights should be booked well in advance (certainly before the holiday starts) since flights are often full.

STAYING IN THE CANARY ISLANDS

Accommodation

More than a million people choose the Canary Islands for their holiday each year, and the figure is still slowly increasing. This has resulted in a very wide choice of accommodation on the more popular islands of Tenerife, Gran Canaria and Lanzarote. In Fuerteventura tourism is now expanding rapidly and a wider range of accommodation is becoming available. Most birdwatchers who visit the islands have accommodation as part of a package holiday, but for those who wish to find their own accommodation, the following hints may be useful.

Hotels

There is little tourism and limited accommodation on La Palma, La Gomera and El Hierro. These three islands all have a Parador (state-run hotel), and these are of a high standard. La Gomera also has the recently opened Hotel Tecina (4 star) at Playa Santiago, which is the best on the island.

In Fuerteventura there is a good range of hotels and apartments both in the north at Corralejo, in the south at Costa Calma and Moro Jable, and at smaller coastal resorts such as Caleta de Fustes (El Castillo). This last resort is very central, and is perhaps the most suitable location for birdwatchers wanting to cover the whole of the island.

For visitors on a lower budget the 'pensions' found in most towns and villages on the larger islands are good value for money. They do vary a little in price but the average is about 2,000 pesetas per night (1996). They are usually kept to a high standard of cleanliness, but do not expect a private bathroom. Pensions are an accommodation option on any of the islands but be prepared to ask for directions as many are away from the main roads. In Fuerteventura the pensions are mainly in Puerto del Rosario.

Camping

The final option is camping. The only official campsites are on Tenerife and Gran Canaria. The addresses are as follows:

Camping Nauta, Canada Blanca (between Guargacho and Guaza), Arona, Tenerife. (Tel: 78-51-18).

Camping Sanssouci, Adeje, Tenerife. (Tel: 78-03-34).

Camping Guantanamo, La Playa de Tauro, Puerto Rico, Gran Canaria. (Tel: 24-17-01).

Camping Temisas, Lomo de la Cruz (Aguimes to San Bartolome), Gran Canaria. (Tel: 79-81-49).

There are no registered sites on any of the other islands but 'off-site' camping is tolerated in many areas. In Fuerteventura, the area of coast just to the north of El Cotillo is commonly used for camping. Please remember, if camping away from a site, to take great care with fire on the wooded islands, and remove all signs of your stay when you leave.

Food

On the main tourist islands there is every kind of restaurant. The islands with less tourism, however, also have a good range of bars and restaurants providing a wide choice, from tapas (Spanish snacks) to full meals. If you want English-style breakfast or any typical English meals, don't head for the local bars. Most restaurants have menus printed in several languages, even in the more out of the way places, so ordering food is not normally a problem. However, the following words will help where necessary:

potaje (de verduras)	–	a thick soup (vegetables)
sopa	–	soup (usually chicken or fish)
bistec	–	steak
chuleta	–	chop
carne	–	meat
cerdo	–	pork
cordero	–	lamb
conejo	–	rabbit
cabra	–	goat
pollo	–	chicken
calamares	–	squid
pulpo	–	octopus
ensalada	–	salad
ensaladilla	–	potato salad with vegetables
higado	–	liver
pescado	–	fish
garbanzas	–	chickpea stew
rancho	–	vegetable and meat stew
carne con papas	–	meat and potato stew
papas arrugadas	–	potatoes boiled in their skins in salted water
papas fritas	–	chips
albondigas	–	meat balls
cerveza	–	beer
vino	–	wine
cafe	–	coffee
con leche	–	with milk
agua	–	water
con gas	–	sparkling
sin gas	–	still
bocadillo	–	a crusty roll
de jamon	–	with ham
de queso	–	with cheese
la cuenta	–	the bill

Banks

Banks are open from 09.00 to 14.00, Monday to Friday (including airport branches). Outside these hours there are exchange offices in the tourist areas. Otherwise, use the hotel reception desks. In all of these, money is changed at the official rate, but the commission charges may vary considerably from one place to another. Cash point machines are fairly widespread and accept Visa and Mastercard.

Shopping

For those people wishing to bring back their duty free allowance of alcohol or tobacco it is recommended to purchase these in the local supermarket, where prices are generally lower than at the airport. Items such as cameras, hi-fi equipment and electronic goods used to be cheaper in the islands than elsewhere, but this is no longer the case.

CLIMATE AND CLOTHING

The Canary Islands are often referred to as 'the Islands of Eternal Springtime' because of their year-round sunny and warm weather. However, the climate is not entirely uniform. The islands have a warm, temperate climate which is relatively mild for their latitude. The main reasons for this relative coolness are the cold sea (influenced by the Canary Current) and the predominantly northeasterly winds.

The Eastern Islands (Lanzarote and Fuerteventura) are drier than the other islands, largely as a result of the lack of high mountains. The Eastern Islands are also rather warmer in summer, and can be hot during periods of easterly winds which can occur at virtually any time of year.

Between 600m and 1,700m there is an area of temperature inversion formed by the sea cooling the air at low levels. Where mountains rise into this zone (this happens on all islands except the two eastern ones, a bank of cloud often forms. This occurs mainly along the northern sides of the islands, where the prevailing winds are deflected upwards by the mountains. This means that the northern parts of these islands are often duller and cooler than the southern sides. The laurel forest occurs in this cloud zone. When such cloud is present, it can severely hinder efforts to find the endemic pigeons and the Blue Chaffinch at its lower sites.

Rainfall is mainly confined to winter months (November-March) but it varies greatly with altitude and location. For example, on Tenerife the average annual rainfall for the high parts of the Anaga peninsula is more than 800mm (31 inches), whereas in the extreme south of the island it is less than 100mm (4 inches). The average temperatures for the coastal region of Tenerife vary from 17.8°C in January (10.5°C minimum, 26.5°C maximum) to 24.2°C in August (17.2°C minimum, 40.4°C maximum). Obviously, there is a temperature gradient with altitude and Mount Teide, on Tenerife, is snow-covered in winter. Snow storms occasionally block the highest roads in winter, although they are soon cleared again.

In the Eastern Islands and in the coastal regions of the other islands, shorts and T-shirts can be worn all year round. However, warmer clothing is often required on the eastern islands in winter during the frequent periods of strong winds. Excursions into the higher mountains of the Western Islands will often require warm trousers, a sweater and a waterproof jacket. As for footwear, a good pair of walking boots or shoes is recommended, although a pair of training shoes will be adequate in many places. An important consideration is that the ground is often very rough, particularly on the plains of the Eastern Islands, so shoes should have good thick soles.

Finally, it is important to remember that the sun is very strong here, especially at higher altitudes. Do not forget the usual protection such as sun block, a wide-brimmed hat and sunglasses. The relatively low temperatures can mislead you into thinking that the sun is not strong enough to burn, but you can be sure that it is. Take particular care when at altitude, and when a strong wind is blowing. Untanned skin can become burnt in half an hour under these conditions.

HEALTH AND MEDICAL FACILITIES

The chance of catching a serious illness in the Canary Islands is very low and common sense will help to ensure that you remain healthy and enjoy your stay.

The most common complaints among visiting tourists are stomach upsets and sunburn/sunstroke. The risk of stomach upsets is reduced by drinking only bottled water, while travellers who are prone to stomach upsets should bring supplies of appropriate medicines as these might not be available from chemists on the islands.

Sunstroke can be avoided by drinking plenty of fluids (but not alcoholic drinks), and it is important to take plenty of water with you on longer excursions. Although bulky, a large thermos flask is ideal for this purpose since it keeps water cool throughout the day. Also, remember that the body loses more water when the skin is exposed than it does when it is covered. Loose-fitting trousers and long-sleeved shirts are therefore better than shorts and T-shirts. For hints on avoiding sunburn see 'Climate and Clothing' (page 10).

There are no dangerous wild animals on any of the islands except for Black Widow Spiders and scorpions, both of which are rare and not often encountered. There are no snakes.

The most likely cause of injury from animals is a dog bite. In rural areas, large dogs are often kept to protect property, and these should be treated with respect. If you are unlucky enough to be bitten, immediate medical attention should be sought.

Tenerife and Gran Canaria have very good medical facilities in all areas. Lanzarote is almost as good, but the other islands are a little more primitive in this respect, and medical facilities are few and far between. Visitors from the UK should take form E111, available from the Department of Health, as this entitles you to reciprocal medical treatment. All visitors should make sure that they have adequate medical insurance to cover all eventualities.

MAPS

There are many maps available for the Canary Islands, most of which are perfectly adequate for birdwatchers. The following is a list of maps available from Stanfords, 12-14 Long Acre, London, WC2E 9LP.

Maps including all the larger islands:

Canary Islands 1:200,000 (Michelin)
Canary Islands 1:150,000 (Clyde Leisure)
T32 - Islas Canarias 1:150,000 (Firestone)

Maps of the Eastern Islands:

Las Palmas 1:200,000 (Spanish National Survey) Provincial Map
Gran Canaria 1:150,000 (Lairs)
Gran Canaria 1:150,000 (Daily Telegraph)
V2 Gran Canaria 1:150,000 (Firestone)
Gran Canaria 1:100,000 (Hildebrand)

Maps of the Western Islands:

Santa Cruz de Tenerife 1:200,000 (Spanish National Survey) Provincial Map
Tenerife 1:150,000 (Mairs)
Tenerife 1:150,000 (Daily Telegraph)
Tenerife 1:100,000 (Hildebrand)
V1 Tenerife 1:150,000 (Firestone)

Also available:

E51 Gran Canaria 1:150,000 (Firestone) Gran Canaria only
E50 Tenerife 1:150,000 (Firestone) Tenerife only
Tenerife 1:125,000 (AA) Tenerife only
Spanish Military Topographic Survey 1:200,000/1:100,000/1:50,000

The maps most widely available in the Canary Islands are T32, E50 and E51, all published by Firestone. The T32 is normally available in all of the island airports as well as many tourist shops. It is not totally up to date, but it a good working map covering all the islands.

WHEN TO GO

The best time for a birdwatching holiday to the Eastern Islands is in the spring. This is the time when resident birds are supplemented by periodic influxes of north-bound migrants. The best period is mid-March to early May, although migrants arrive a little earlier in some years. It must be said, however, that reasonable numbers of migrants only occur during periods of southeasterly or easterly winds. Also, it should be noted that Eleonora's Falcon does not return until later in May. March and April are also the months when most of the resident species are at the height of their breeding season.

Although migrants also pass through the Eastern Islands in the autumn, the limited observations which have been made to date suggest that both the range of species and the numbers involved are lower

The best time to visit the other islands is from early May until the end of August. Although the endemic species can be seen at any time of year, the majority of the rarer seabirds can only be seen during this period. Overall, the best time is probably either May or August.

If cost is an important consideration the Christmas and Easter periods should be avoided, since prices are generally higher at these times. Prices are also higher during mid-summer, and tend to be low in February/March.

No matter what time of year the islands are visited, it is possible to find something totally unexpected if you look hard enough.

Plain Swifts

INTRODUCTION TO THE SITE INFORMATION

In the following site information, each island is dealt with separately, and the sites are slightly biased towards finding the endemic or special species. Sites are included for all the main islands, although the more important birdwatching islands (Tenerife, La Gomera and Fuerteventura) are covered in greater detail. Sites for three of the four ecological zones of the islands are included. The three zones included are the lower arid zone (often characterised by succulent Euphorbia species); the Laurel Forest (Monteverde) with its multitude of endemic plants; and the Pine Forest (Pinar). The only zone not included is the High Mountain zone, which includes a sub-alpine element in the flora but is of little interest ornithologically.

The species listed for each site are the endemics which are found there and some of the commoner species encountered. Sketch maps are included for some of the areas covered, but the directions given should be detailed enough to find all the sites.

TENERIFE

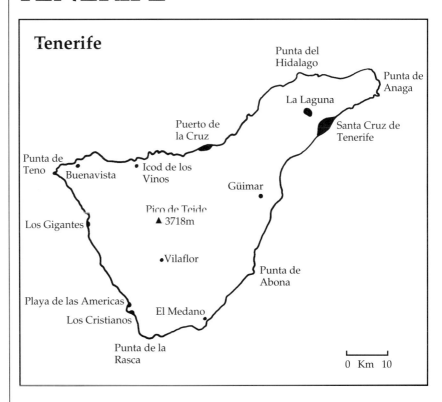

Tenerife holds four Canarian endemics – Bolle's Pigeon, Laurel Pigeon, Tenerife Goldcrest and Blue Chaffinch; two Macaronesian endemics – Berthelot's Pipit and Canary; and one species which is largely confined to Macaronesia – Plain Swift. There are also several interesting seabirds.

The following selection of sites should give any visiting birdwatcher the maximum opportunity of seeing all the endemics and hopefully some of the seabirds as well. All the sites in a given region are listed together. The regions are south, northwest, northeast, east, and central (around Mt. Teide).

The South

El Médano (1)

El Médano is a village with a couple of hotels and apartment complexes, which is situated on the only natural pale-sand beach on Tenerife. The bays here can be good for migrating and wintering waders as can the salt water pool at the base of Montaña Roja. These bays are also one of the few remaining breeding sites on Tenerife for Kentish Plover.

Location El Médano is on the south coast, northeast of the airport and is reached by turning off the motorway at the El Médano/San Isidro exit and heading for the coast.

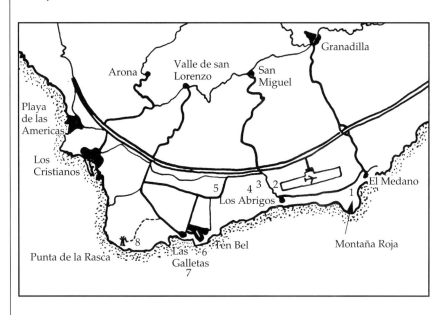

The bays between the harbour wall and the new Hotel Windsurf can be viewed from roads. The bay between the village and Montaña Roja can be viewed either from the village, from above the pool or by walking along it. On the western side of Montaña Roja is another bay worth looking over, but be careful where you aim your optical equipment as part of this beach is reserved for nudists. A gull roost can be found at all times of the year either on this western bay or on the disused airstrip to the north of Montaña Roja, usually early in the morning.

In addition to the birds mentioned above, other species in the area include Kestrel, Lesser Black-backed Gull (winter), Yellow-legged Gull, Plain Swift (summer), Hoopoe, Lesser Short-toed Lark (very rare), Berthelot's Pipit, Great Grey Shrike, Spectacled Warbler, Spanish Sparrow, Linnet and Trumpeter Finch (uncommon). Migrant waders include Ringed and Grey Plovers, Knot, Sanderling, Little Stint, Curlew Sandpiper, Dunlin, Ruff, Black-tailed and Bar-tailed Godwits, Whimbrel, Curlew (rare), Redshank, Greenshank, Common Sandpiper, and Turnstone.

Rarities which have been recorded in the area include Black Kite, Hobby, Barbary Falcon, Oystercatcher, Black-winged Stilt, Avocet, Cream-coloured Courser, Temminck's Stint, White-rumped Sandpiper, Lesser Yellowlegs, Audouin's, Ring-billed, Common and Greater Black-backed Gulls.

Embalse de Ciguaña (2)

This is one of many disused dams, and hence only contains water after rain. Most years the winter provides enough rain to leave water until September or October, but some years it can be totally dry before then. This is a site where Barbary Partridges can be seen as they occasionally come here to drink, as do Trumpeter Finches. Little Ringed

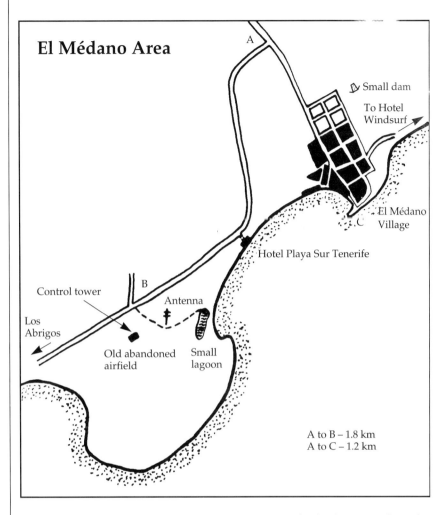

Plovers nest here. The dam also attracts herons, ducks (in winter), and waders (on passage).

Location

One kilometre northwest of Los Abrigos, on the eastern side of the road, is a small red and white striped building with a red antenna and surrounded by green chain-link fencing. Next to this building there is a small pull-in just large enough for a couple of vehicles. Park here and follow the track uphill for about 200m to the wall of the small dam.

Birds

Some of the other species recorded here include Grey Heron, Teal, Kentish, Ringed and Grey Plovers, Sanderling, Snipe, Black-tailed Godwit, Green, Wood and Common Sandpipers, and Grey and White Wagtails. Rare species which have been recorded here are White-rumped, Pectoral and Buff-breasted Sandpipers.

Golf del Sur (3)

This golf course is an area of lush vegetation surrounded by rocky, volcanic desert scrub. A walk or drive around the whole complex can be rewarding as the course is a magnet for birds.

Location   Continue on from Embalse de Ciguaña for 2km, or leave the motorway at the San Miguel exit and take the road towards Los Abrigos, to reach the entrance to Golf del Sur golf course.

Strategy   The most productive areas appear to be the fairways on either side of the road before the club house. Please remember that the golfers have complete priority, and birdwatchers should never get in their way. It is also wise to keep a sharp eye open for stray shots!

Birds   Regular and resident species in the area include Kestrel, Stone-curlew, Little Ringed Plover, Lesser Black-backed Gull (winter), Yellow-legged Gull, Turtle Dove (summer), Barn Owl, Plain Swift, Hoopoe, Berthelot's Pipit, Grey Wagtail, Great Grey Shrike, Spectacled Warbler

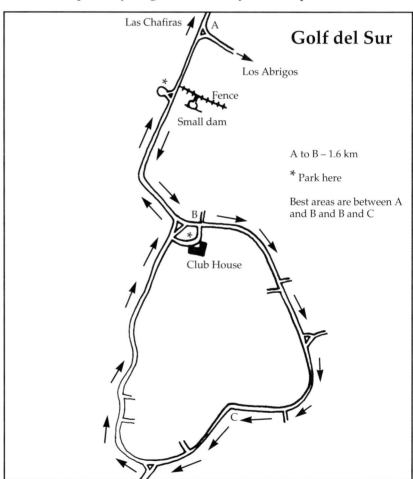

Las Chafiras — A

Golf del Sur

Los Abrigos

Fence

Small dam

A to B – 1.6 km

* Park here

Best areas are between A and B and B and C

B

*

Club House

C

and Trumpeter Finch. Rare visitors from the surrounding desert scrub and deserted agricultural land are Barbary Partridge and Lesser Short-toed Lark.

In recent years the golf course has proved to be a fairly regular wintering site for Red-throated Pipit – usually about 5-10 birds but, occasionally, as many as 20 or more have been seen in a single flock.

Skylark and Meadow Pipit are also irregular winter visitors, but White Wagtail is annual and fairly common. Waders are often encountered on the greens and fairways, especially during spring and autumn migration. Some species such as Whimbrel, and Grey and Ringed Plovers remain throughout the winter. The small dam on the left at the entrance to the golf course can be quite productive if the water level is not too high.

The whole development is good for rarer birds and over the last few years a good variety of species has been recorded. These have included Cattle and Little Egrets, Grey Heron, Spoonbill, Marsh Harrier, Merlin, Eleonora's and Barbary Falcons, Little Crake, Collared Pratincole (annual in late spring), American Golden Plover, Little Stint, Pectoral, Curlew, Buff-breasted and Wood Sandpipers, Whiskered Tern, Pallid Swift, Bee-eater, Red-rumped Swallow, Short-toed Lark, Richard's and Tawny Pipits, Woodchat Shrike, Starling (winter), Sedge and Subalpine Warblers, and Whinchat.

Amarilla Golf and Country Club (4)

This golf course is not as popular as Golf del Sur because it is not in such good condition. However, this makes it better for birds. It is not unknown for birds flushed from Golf del Sur by golfers to spend most of their day on Amarilla Golf.

Location

Bordering on the western side of Golf del Sur is Amarilla Golf and Country Club.

Strategy

Probably the best area to search for birds is the three holes (and the bare ground enclosed by these) on the left on the approach to the club house. If stopped by the security guards, tell them you are going to the club house. Park in the car park at the club house. Other parts of the course can be viewed from the main road through the development down to the coast. Please remember that the golfers have complete priority, and birdwatchers should never get in their way. Remember to keep a sharp eye open for stray shots!

The dam on the left just before the guard-post occasionally attracts a few waders and both Spoonbill and Little Egret have been recorded.

Birds

The species around Amarilla Golf are much the same as on the neighbouring Golf del Sur but both Trumpeter Finch and Lesser Short-toed Lark are recorded regularly here. The list of rarities for this site includes Glossy Ibis, Barbary Falcon, Cream-coloured Courser, Dotterel, American Golden Plover, Baird's and Butt-breasted Sandpipers, Red-throated Pipit, Woodchat Shrike and Isabelline Wheatear.

Guargacho 1, 2 and 3 (5)

These three dams, situated around the village of Guargacho, are only two or three kilometres from Amarilla Golf and Country Club. Guargacho 1 usually has some water in it all year round and attracts a variety of birds. However, dams 2 and 3 are dry, except after heavy rain. The desert scrub and hills to the east of the first dam hold a small population of Barbary Partridge, but they can be very difficult to find.

Location   From the San Miguel exit of the motorway take the road to Las Galletas, pass the entrance to Amarilla then turn left towards Las Galletas. After 1.6km is the village of Guargacho and the first dam can be clearly seen on the left of the road, opposite the small church.

Birds   Resident species include Hoopoe, Berthelot's Pipit, Grey Wagtail, Spectacled Warbler and Spanish Sparrow. Rock Sparrow and Trumpeter Finch used to drink at Guargacho 1 a few times a day but now their visits are much less frequent as both species seem to be declining in the area. Rock Sparrow can sometimes be seen on the overhead wires or television antennae around the village. Little Ringed Plover breeds and Moorhen has bred at Guargacho 1. As with any of the other dams on Tenerife unusual species and rarities turn up fairly regularly, and have included Cattle Egret (annual in winter), Night and Purple Herons, Glossy Ibis, Shelduck, Blue-winged Teal, Barbary Falcon, Spotted, Little and Baillon's Crakes, Black-winged Stilt, Little Stint, Curlew Sandpiper, Black-tailed Godwit, Marsh Sandpiper, Red-rumped Swallow, Tree and Tawny Pipits, Woodchat Shrike and Bluethroat.

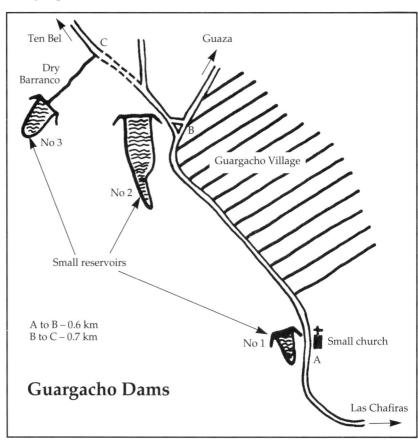

Ten Bel
C
Guaza
Dry Barranco
No 3
B
Guargacho Village
No 2
Small reservoirs
A to B – 0.6 km
B to C – 0.7 km
No 1
Small church
A

Guargacho Dams

Las Chafiras

Ten-Bel (6)

This holiday complex contains the only large number of trees in the coastal region between El Médano and Los Cristianos/Las Amíricas. The area acts as a migrant trap under the correct weather conditions.

Location   It is situated to the east of Las Galletas, and is classified as part of Costa del Silencio.

Strategy   The best time to visit this site is in the spring or autumn, during an east or southeast wind, preferably associated with a Saharan dust storm.

Birds   Resident species include Hoopoe, Blackcap, Spectacled Warbler, Blackbird and Spanish Sparrow. Both Greenfinch and Starling occur occasionally in winter. Some of the rarer species recorded here include Blue-cheeked Bee-eater, Bee-eater, Woodchat Shrike, Subalpine Warbler, Pied Flycatcher and Serin.

Los Palos

This is a small pitch and putt golf course and golf training centre which can be good for migrants

Location   The course lies between Guaza and Las Galletas.

Strategy   The site is best visited during the early morning or late afternoon when there are no golfers playing on the course.

Birds   Hoopoe, Berthelot's Pipit and Great Grey Shrike are resident. The following migrants have been recorded: Cattle Egret, Ruff, Skylark, Red-throated Pipit (winter), Yellow Wagtail, White Wagtail (winter), Whinchat, and Wheatear.

Las Galletas (7)

This coastal village has a small harbour and a rocky reef which is exposed at low tide. The harbour normally holds very little bird life except for Yellow-legged Gull, although Black Tern and Little Gull have been seen here. The rocks and reef area, in contrast, attract a reasonable variety of species, except in summer.

Location   The village lies west of Ten-Bel, the harbour is at the western end and the reef is at the eastern end.

Strategy   Visit the area at low tide when the reef is used as a roosting site by terns and other species.

Birds   Species which can be seen here are Little Egret, Grey Plover, Whimbrel, Turnstone, Yellow-legged Gull, and, depending on the state of the tide, Sandwich Tern. The terns are most numerous on passage, with up to 100 roosting. An unidentified, large red-billed tern has been seen here, so it is always worth looking carefully through the flock.

Punta de la Rasca and Roquito del Fraile (8)

Punta de la Rasca is a headland with a lighthouse which can be a good seawatching spot. The surrounding abandoned fields sometimes attract migrants and there is a irregular roosting site for Long-eared Owl here. Roquito del Fraile is a reservoir which regularly attracts waterbirds and has produced quite a few rarities, including a number of first records for the Canary Islands.

Location

Coming from Los Cristianos/Las Amìricas along the motorway, take the Las Galletas exit and drive through Guaza. Look for a Repsol petrol station on the right, about 4.3km from the motorway. If coming from Las Galletas the petrol station is on the left, after approximately 2.5km. Take the dirt track between the banana plantation wall and the petrol station. At the end of this track turn right and continue to the end of the plantation walls where it is possible to park just before a red and white barrier. Please park with consideration as these tracks are used by the local people and large trucks pass along them on occasions. From here, continue on foot past the fairly new chalet to the fork in the track and then bear left to the large barn with a line of Indian Laurel trees behind. This track leads to a locked gate where it becomes a tarmac road going to the lighthouse, Faro de la Rasca, which is the best seawatching spot.

About 200m along the road from the locked gate towards the lighthouse, on the first right bend, a track leads off to the left towards some small houses, a large tomato shed and a large concrete fence with large circular holes in it. The reservoir known as Roquito del Fraile is located inside this concrete fence.

Strategy

From the lighthouse there is a chance of seeing some of the rarer breeding seabirds, although a lot of time and patience and a telescope is likely to be required.

The abandoned agricultural fields between the locked gate and the lighthouse used to be a good area for Barbary Partridge and Lesser Short-toed Lark, but these are now very rare in the area. There is still a small population of Trumpeter Finches as well as Stone-curlew, Hoopoe, Plain Swift, Berthelot's Pipit, Great Grey Shrike and Spectacled Warbler here. This is almost certainly the best place in Tenerife to look for migrants, although the birds can turn up anywhere due to the lack of good migrant habitat. The vegetation here (Euphorbia scrub) does not encourage birds to stay for very long. However, regular watching can produce rewards, especially in southeasterly Sirocco weather conditions.

Approach the reservoir quietly and peer through the circular holes in the wall. Some of the birds may be quite close to you.

Birds

Over the sea Cory's Shearwaters are usually abundant and species such as Bulwer's Petrel, Little Shearwater, British Storm-petrel and Madeiran Storm-petrel have been seen.

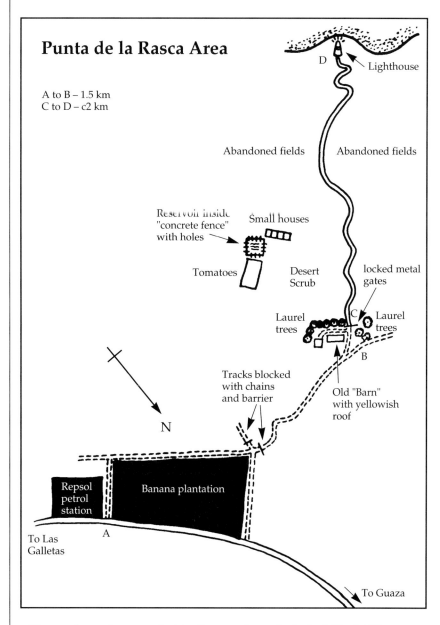

Punta de la Rasca Area

A to B – 1.5 km
C to D – c2 km

D — Lighthouse

Abandoned fields   Abandoned fields

Reservoir inside
"concrete fence"
with holes

Small houses

Tomatoes

Desert
Scrub

locked metal
gates

Laurel
trees

C   Laurel
trees

B

Tracks blocked
with chains
and barrier

Old "Barn"
with yellowish
roof

N

Repsol
petrol
station

Banana plantation

To Las
Galletas

A

To Guaza

Unusual species recorded in the area have included Black Kite, Marsh, Hen and Montagu's Harriers, Eleonora's and Barbary Falcons, Quail, Pallid, Alpine Swift, White-rumped and Little Swifts, Bee-eater, Red-rumped Swallow, Short-toed Lark, Tawny, Tree and Red-throated Pipits, Yellow Wagtail, Woodchat Shrike, Olivaceous, Orphean, Subalpine and Willow Warblers, Pied Flycatcher, Wheatear, Black-eared Wheatear, and Desert Wheatear. Perhaps the local speciality of this area is Long-eared Owl which can sometimes be found roosting in the laurel trees behind the barn, or in the isolated laurel tree to the right of the locked gate.

The reservoir is one of the few breeding sites on Tenerife for Coot, and attracts a large variety of waders and waterfowl during migration and in winter. In summer there is usually the occasional wader, and Grey Heron and Little Egret are regular visitors. This is one of the best birding sites for rarities in Tenerife. Some of the unusual birds recorded here include Black-necked Grebe, Cormorant, Cattle and Great White Egrets, White Stork, Spoonbill, White-fronted, Greylag and Brent Geese, Wigeon, Gadwall, Green-winged Teal, Mallard, Garganey, Ring-necked Duck, Lesser Scaup, Peregrine Falcon, Black-winged Stilt, Avocet, Collared Pratincole, American Golden Plover, Temminck's Stint, Baird's, Pectoral and Curlew Sandpipers, Long-billed Dowitcher, Little, Sabine's and Common Gulls, Whiskered and Black Terns, Spotted Flycatcher, Serin and Snow Bunting.

Cory's Shearwaters and Pilot Whales

The Northwest

Arguayo (1)

This village, southeast of Santiago del Teide, is reached by taking the minor road from Chio to Santiago del Teide. The electricity wires around the village should be checked for Rock Sparrow, which can often be seen here.

Santiago del Teide (2)

The area on the south side of the village is another good area for Rock Sparrow. Check the fields and wires.

Erjos Ponds (3)

These ponds are found just off the main road between Santiago del Teide and Garachico. After Santiago del Teide the road climbs to Puerto de Erjos at 1117m and then descends to Restaurante Fleitas, which is just after a turning on the right to Los Llanos. One kilometre further on

Location

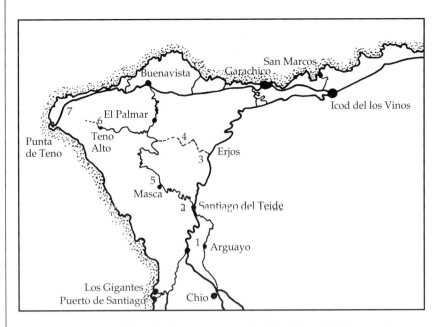

is a minor crossroad with a bus stop on the left. Turn left here and follow this track, past a couple of houses, to its end where the Erjos Ponds can be seen.

Birds   These ponds are one of the few breeding sites on Tenerife for Moorhen, and are also an excellent area for Sardinian Warbler and Canary. Chiffchaff and Linnet are probably the most common species, but other regular species are Grey Heron, Sparrowhawk, Buzzard, Kestrel, Yellow-legged Gull, Rock Dove, Turtle Dove (summer), Plain Swift, Berthelot's Pipit, Grey Wagtail, Robin, Blackbird, Spectacled Warbler, Blackcap, Blue Tit, Raven, Spanish Sparrow, Chaffinch, Greenfinch, Goldfinch and Corn Bunting. Common Sandpiper is regular in winter. Swallow and House Martin are regular on passage, especially in spring. Osprey is an occasional visitor. Accidental species have included Glossy Ibis, Night and Purple Herons, Green-winged Teal, Ring-necked Duck, Barbary Falcon and Alpine Swift.

Monte del Agua (Erjos) (4)

This is one of the few remaining patches of Laurel Forest on Tenerife and is an important site for the two endemic pigeons which are confined to this habitat.

Location   The access track to this area leads off the Santiago del Teide to Garachico road only, 200m after the turning to the Erjos ponds. The track is on the left just, before the 'casa forestal' and Erjos sign which are on the right.

Strategy   Probably the best vantage point is 4.9km from the start of the rough track where there is a conspicuous rock on the right hand side of the track with a green rain gauge on the top. From the top of the rock and the edge of the track there are excellent views over the forested valley in which the pigeons live although mist can hamper visibility. The two pigeons can both be seen at this site with time and patience but the views are almost always of birds in flight, although they do sometimes pass very close by. The track is not in a very good condition and a great deal of care should be taken when driving along it. From the rock it is possible to continue for a further 6.2km to its end near El Palmar, or to return to the road through Erjos to Garachico.

It is always worth spending a few minutes looking around the fields at the start of the track and the garden of the 'casa forestal' as these are both good places for Canary.

Monte del Agua

Birds   Other species which can be seen in and around the laurel forest are Sparrowhawk, Buzzard, Kestrel, Barbary Partridge, Woodcock, Rock Dove, Turtle Dove (summer), Barn and Long-eared Owls, Plain Swift, Berthelot's Pipit, Grey Wagtail, Robin, Blackbird, Spectacled and Sardinian Warblers, Blackcap, Chiffchaff, Tenerife Goldcrest, Blue Tit, Raven, Spanish Sparrow, Chaffinch, Canary, and Linnet. Accidental species are seldom found here but species such as Eleonora's Falcon, Woodpigeon, Alpine Swift, Melodious Warbler and Willow Warbler have all been recorded.

Masca (5)

This is a small mountain village set in some of the most spectacular scenery on Tenerife. The beautiful views attract many ordinary tourists but birdwatchers can also find the surrounding fields and barrancos (dry gullies) rewarding. In Santiago del Teide take the turning signposted to Masca and Buenavista. Masca is only about 6km from Santiago del Teide, but the access road is particularly windy and great care should be taken when driving in and out of this area. Masca is another site in the northwest of Tenerife where Rock Sparrow can be found. Barbary Partridge also occurs here, although they can be very difficult to find. The area is good for Kestrel and Buzzard.

Teno Alto (6)

Take the minor road signposted to the left, off the Masca to Buenavista road south-southwest of El Palmar. The area around the village can be good for Rock Sparrow but they are sporadic in their appearance.

Punta de Teno (7)

The northwestern tip of Tenerife is reached by taking the minor road west from Buenavista. This area is seldom visited by birdwatchers, but must surely be the best seawatching spot on the island. Unfortunately, at the time of writing, the point is closed to the public, but the local police say that once the renovations to the lighthouse have been completed access to the point will be restored. There are two main habitats on the headland – Euphorbia scrub and sea cliffs – neither of which have a lot of bird species, though there are a few local specialities The cliffs on the north coast are a reliable site for Barbary Falcon.

Strategy   From Buenavista take the left turn signposted to Parque Natural Punta de Teno. After 4.1km, where there is an arch of rock over the road, there is a small lookout from which Barbary Falcon can sometimes be seen. From the lookout, scan the cliffs to the west, where the birds perch on cliff ledges. Continue along the road and pass through a 600m long tunnel. After another 400m, just past a left hand bend, there is a very small lay-by on the right which is only large enough for one car. Park here and walk back to the bend to view the cliff-face. Barbary Falcons are regularly seen on the ledges of this large cliff. The last area to look for the falcon is another 200m further on, where there is more room to park off the road and scan the cliffs. After this the road starts to descend, leaves the cliff behind and after 2.9km passes through euphorbia scrub with some tomato plantations to the right of the road. It is here, in winter, that a flock of about 300 Rock Sparrows can be found. The road continues to the point, 9.4km from Buenavista, where a seawatch is always worthwhile.

Birds   A few Canaries, and a mixed flock of Skylarks and Lesser Short-toed Larks can be found in the area, as well as commoner species such as

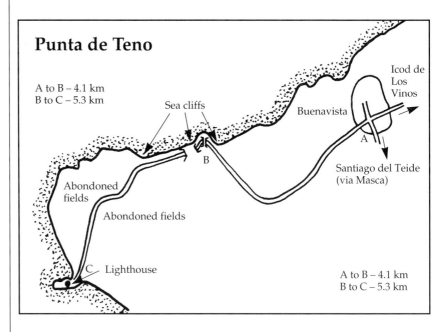

Berthelot's Pipit, Spectacled Warbler, Raven, Spanish Sparrow and Linnet. With luck, Barbary Partridge or Trumpeter Finch may be seen.

Over the sea Cory's Shearwaters are abundant, and other species recorded include Bulwer's Petrel, Manx and Little Shearwaters, and British Storm-petrel. Accidental species recorded from here include Red-billed Tropicbird, and Great and Pomarine Skuas.

The North

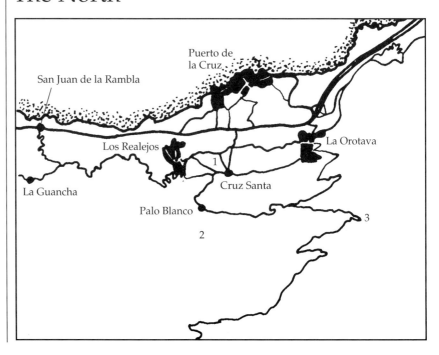

Los Realejos Reservoir
(also known as Embalse de la Cruz Santa) (1)

From Los Realejos take the TF-212 towards Cruz Santa and La Orotava. After 1.3km bear left at a mini-roundabout and view the reservoir from here. The reservoir occasionally attracts ducks such as Teal in winter, and is always worth a quick look when passing by. In the winter of 1995-96 two Ring-billed Gulls were regularly observed at this site.

Chanajiga (2)

This isolated area of laurel forest is possibly the best site on Tenerife for the two species of endemic pigeon, but it can be severely affected by

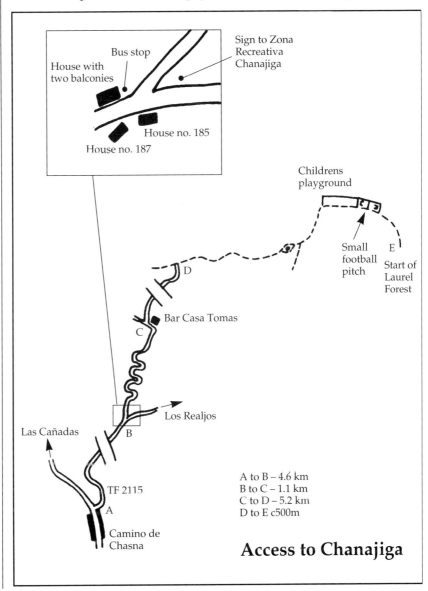

Access to Chanajiga

adverse weather conditions, especially low cloud. Access is much more complicated than at Monte del Agua, so the directions given here must be followed very carefully in order to avoid getting lost.

Location

From La Orotava take the road towards Las Cañadas and upon reaching the village of Camino de Chasna, after about 8.5km, take the right turn on TF 2115 towards Palo Blanco and Los Realejos. 4.6km along this road is a left turn which is unsignposted coming from this direction. However, if approaching from Los Realejos there is a sign just before the turning to Las Llanadas and Zona Recreativa Chanajiga.

Take this minor road through the village of Los Llanadas (there is no village name-plate). At the Casa Tomas bar take a ninety degree left hand bend and then take the main right fork. Continue for another 5.2km and the tarmac road terminates at a T-junction. Turn right here, follow the track through the picnic area and the children's playground, and pass the football pitch to reach the laurel forest.

Strategy

The area is frequently shrouded in mist and low cloud, and it is often raining, so be prepared for the worst.

Birds

As well as the two pigeons, species found here include: Sparrowhawk, Buzzard, Kestrel, Woodcock, Rock Dove, Turtle Dove (summer), Long-eared Owl, Plain Swift, Berthelot's Pipit, Grey Wagtail, Robin, Blackbird, Sardinian Warbler, Blackcap, Chiffchaff, Tenerife Goldcrest, Blue Tit, Raven, Chaffinch, Canary, and Linnet.

Aguamansa (3)

This is a very unusual site because Chaffinch and Blue Chaffinch can be found side by side.

Location

From La Orotava take the road to Las Cañadas and after about 9km (about 2km passed the village of Aguamansa) is a left turn to La Caldera. On reaching the end of this road stop in the parking area.

Strategy

The habitat is a mixture of pine, laurel and tree heath, a combination suitable for sustaining both species of chaffinch. Follow the minor road which encircles the picnic area to search for Blue Chaffinch. The trees and bushes by the bar often produce both species. This area also contains many of the species found at Chanajiga and is especially good for Sparrowhawk.

The Northeast

Dársena Pesquera (Fish Quay) (1)

It is always worth a quick look here at the gulls, when passing en-route to the Anaga Peninsula. Leave Santa Cruz on the coast road towards Playa de Las Teresitas and after about 5km follow the signs to Dársena Pesquera. It occasionally attracts Sandwich Tern and Common

Gull in winter, and Common and Arctic Terns on passage. It is also the place where the first Bar-tailed Desert Lark for the Canary Islands was found.

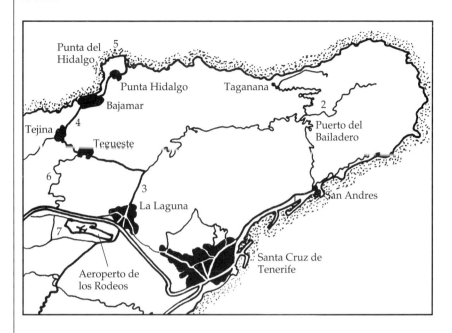

Anaga Peninsula (2)

This is a good area for obtaining perched views of Bolle's Pigeon, which can often only be seen in flight elsewhere.

Location
Continue from the Dársena Pesquera to San Andrís. Turn left there onto the TF112 towards El Bailadero and Taganana. Ignore the turning to Taganana (TF1134) and continue on the TF1123. After 2km turn right onto the TF1122 towards El Bailadero and Chamorga. It is only about 450m to the bar at El Bailadero and then another 1.5km to El Pijaral. Watch carefully for a sign saying Monte U.P.no. 46, San Andrís, Pijaral, Igueste y Anaga, on the right of the road on a left hand bend. Then, after a right hand bend, there are some steps cut into the bank on the left and signposted to El Pijaral. Park on the right at this point.

Strategy
Walk back to the bends in the road and look over the wooded valley on the southern side of the ridge. Bolle's Pigeon can usually be seen easily, and with a little patience it is possible to get excellent views of the birds perched on exposed branches. This area is often very windy and is frequently rainy or engulfed in cloud. Obviously, this affects the viewing conditions and can make the birds harder to find and less likely to perch in the open. Most of the other woodland species can be seen here or further along the road towards Chamorga.

Continuing from El Pijaral after 2.9km is the Parque Forestal, which is a public picnic/barbecue area. The trees around here usually hold good numbers of Tenerife Goldcrest, but they can be hard to see.

After another 1.5km there is a track on the left signposted to Cabezo

Bolle's Pigeon

del Tejo which, although not as good as El Pijaral, can also provide views of Bolle's Pigeon. The only reason to continue any further along this road is to look for Rock Sparrows around Chamorga, though they are not common. As this is a no-through road, you should turn around at this point and return to the main road to Monte de las Mercedes. Be very careful along the road between El Bailadero and Chamorga as it is very narrow in places and can be surprisingly busy, especially at weekends or in periods of good weather, when the inhabitants of Santa Cruz come out for a drive or a picnic. The road from El Bailadero to Monte de las Mercedes also deserves care as it is very narrow in places.

About 10.5km from El Bailadero the road comes to a T-junction. Turn left here towards Pico del Ingles. After 1km is the parking area for the lookout (mirador). Park here and walk the 50m to the lookout to view a large area of laurel forest on the hillside below. This site is very good for Bolle's Pigeon and you can sometimes see them perched. This is the only site in the northeast where Laurel Pigeon can be seen, but it is very rare here.

La Laguna (3)

This is the easiest and most regular site in Tenerife for Serin.

Location

From Pico del Ingles, take the road towards La Laguna. This road passes Mirador Cruz del Carmen and then the road descends to Las Mercedes and then to a T-junction at Las Canteras. Turn left at the T-junction towards La Laguna and after 1.7km there is a set of traffic lights. Continue on the main road for another 400m and then take the minor road to the left. This road is straight for about 500m, then reaches a crossroads, just before the junction is an access road which can be used for parking.

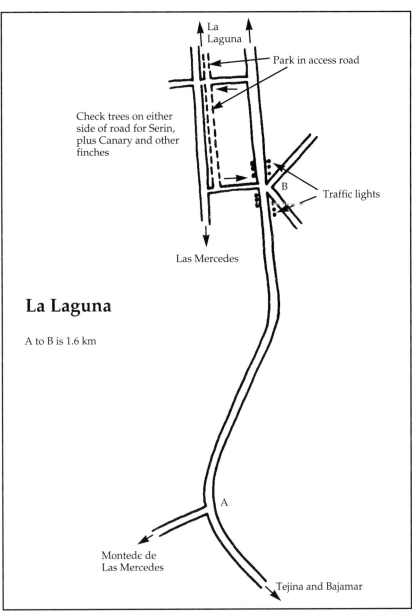

La Laguna

Park in access road

Check trees on either side of road for Serin, plus Canary and other finches

B

Traffic lights

Las Mercedes

La Laguna

A to B is 1.6 km

Montede de Las Mercedes

A

Tejina and Bajamar

Strategy   Walk along the roadside and listen for the characteristic high-pitched jingling song of Serin. Please be very careful at this site as the cars often travel very fast along this major road.

Birds   Species in this area include: Canary, Greenfinch, Goldfinch and Corn Bunting. Starling (a rare and localised breeding bird in Tenerife) might also be seen as there are a few pairs in the area. Near to the Serin area is a small stream with breeding Moorhen and Grey Wagtail.

Tejina Ponds (4)

This area attracts a variety of species during migration and in winter and is an important breeding site for Moorhen.

Location Leave La Laguna heading north, on the road signposted to Bajamar. You will pass through Las Canteras and Tegueste and then reach Tejina. In Tejina follow the signs to Bajamar and Punta Hidalgo. Park on the outskirts of Tejina, where the road crosses a barranco (dry gully) just before the bar El Puente on the right.

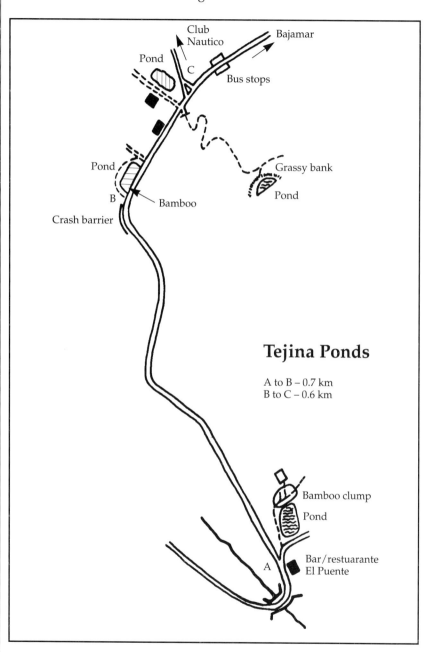

Tejina Ponds

A to B – 0.7 km
B to C – 0.6 km

Strategy Immediately after the bar, also on the right, is a minor road. Walk up this road to the first bend from where a track leads off to the left to a small dam. Walk around this dam and through the vegetation at the far end to a second, much smaller dam.

Return to the car and continue along the main road towards Bajamar. The next dam is 750m from the bar, on the left side of the road. Park carefully on the roadside on the right and be very careful crossing this busy road to view the dam.

About another 350m from this dam there is a left turn signposted to the Club Nautico. Park carefully here and walk back along the main road the short distance to the track leading off on the opposite side. The dam here is inside the grassy bank on the right of the track after about 250m. Opposite the start of this track is a group of three very small dams which do not attract many birds but are worth a quick look just in case.

Birds   Species such as Purple and Night Herons, Little and Cattle Egrets, Squacco Heron, Little Bittern, White Stork, Shoveler, Wigeon, Teal, Scaup, Osprey, Marsh Harrier, Spotted, Little and Baillon's Crakes, Bee-eater, Cuckoo, Red-rumped Swallow, Crag Martin and Sedge Warbler have all been recorded. Coot has attempted to breed and Grey Heron may breed in the area.

Punta Hidalgo (5)

During good weather conditions in spring Punta Hidalgo can be alive with migrants. It is also a good place for seawatching.

Location   The area lies 2km northeast of Bajamar and is easily accessed from the main road.

Strategy   This is quite a large area and most of it is private with no public access. The two main areas to concentrate on are the fields between the main road and the large hotel, and the fields and hedges below the end of the road.

Birds   Many migrants and vagrants have been recorded in the area including Purple Sandpiper, Alpine Swift, Bee-eater, Calandra and Short-toed Larks, Red-rumped Swallow, House and Sand Martins, White Wagtail (winter), Yellow Wagtail, Red-throated Pipit, Woodchat and Great Grey Shrikes, Melodious, Orphean, Subalpine and Wood Warblers, Red-breasted Flycatcher, Black Redstart (winter), Wheatear, and Desert Wheatear.

Valle Molino Reservoir (also known as Embalse de Valle Molina) (6)

This is an important site in the northeast for migrants and winter visitors, but it does not have any special breeding species.

Location   Leave the northern motorway at the exit to Los Rodeos (the airport) and follow the road which runs parallel to the motorway on the northern side towards Guamasa and Tacaronte. After about 1km take the right turn to Tegueste which passes through El Portezuelo. 3.8km along this road, park on the left at the entrance to the Depísito Regulador de Valle Molina, just before a small school. Walk along the entrance road to view the reservoir from the locked gates.

Birds   Species seen regularly around here include Little Egret, Grey Heron, Buzzard, Kestrel, Sparrowhawk, Berthelot's Pipit, Grey Wagtail, Tenerife Goldcrest, Canary and Greenfinch. Migrants in the last few years have included Black-necked Grebe, Tufted Duck, Pochard, Ring-necked Duck, Wigeon, Teal, Shoveler, Spoonbill, Collared Pratincole, Grey Plover, Lapwing, Black-tailed Godwit, Spotted Redshank, Green, and Wood Sandpipers, Common Sandpiper (regular in winter), Black-headed Gull (common in winter) and Siskin. Quail can often be heard in the fields nearby.

Los Rodeos (7)

This is an area of fields and grassland which is good for Lesser Short-toed Lark, Berthelot's Pipit, Linnet and Corn Bunting. Quail can be heard easily in this area in spring, but seeing them is difficult.

Location   Take the road from La Laguna to La Esperanza and Las Cañadas. About 1km beyond the roundabout on the south side of La Laguna, the airport perimeter fence becomes visible on the right. At the end of the fence, after about 500m, there is a road on the right which follows the southern boundary of the airport. The fields on the left of this road, and the grassland inside the airport fence on the right can be very productive for the first 750m and should produce the species mentioned above.

Birds   Other species and accidentals seen in the area include Marsh, Hen, Pallid and Montagu's Harriers, Buzzard, Collared Pratincole, Dotterel, Great Spotted Cuckoo, Short-eared Owl (winter), Short-toed Lark, Tawny, Meadow and Red-throated Pipits, Swallow, House Martin, Woodchat Shrike, Great Reed and Willow Warblers, Black-eared and Desert Wheatears, and Song Thrush.

The East

This is the poorest region in Tenerife for birdwatching and there are only a couple of sites worth visiting which are described only briefly.

La Hidalga (1)

South of Candelaria, take the Arafo exit off the motorway and follow the signs for Arafo. Just before reaching a T-junction there is a small dam on the right. Usually, the dam is totally birdless but it has attracted one or two species of interest over the years, including Little Egret, Grey Heron, Spoonbill, Teal, Moorhen, Coot, Black-winged Stilt, Collared Pratincole, Wood Sandpiper and Red-rumped Swallow.

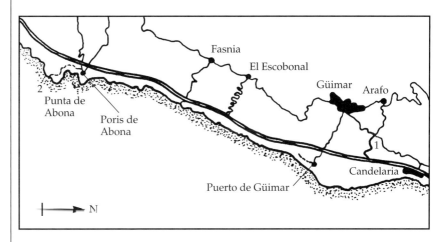

Punta de Abona (2)

This is a good seawatching headland, with easy access, about 20km northeast of the airport at Reina Sofia. Leave the motorway at the Poris de Abona exit and drive into Poris. In the village take the left turn signposted to Punta de Abona and follow this road through the few houses known as La Zarnova down to the lighthouse, 2.4km from Poris. The site is not watched regularly but Little Shearwater has been recorded. Sandwich Tern occurs in winter and the coast can produce a variety of waders including Ringed Plover and Common Sandpiper.

Central

All these sites are at fairly high altitude around Mount Teide, and most are in the pine forest belt, the habitat of the endemic Blue Chaffinch. As they are mainly included as sites for this species, they are described only briefly.

Vilaflor (1)

At 1371m this is the highest town in Tenerife, and it is worth checking the wires for Rock Sparrow, which occurs fairly frequently around the town. 2km from Vilaflor on the road to Boca de Tauce and Mount Teide there is a very large pine tree, with a small car park, on the right. Park here to view the bird table in the 'Casa Forestal' garden on the opposite side of the road, just before the entrance to the car park. Food is put out on the bird table and often attracts Great Spotted Woodpecker, Blue Tit, Blue Chaffinch and Canary.

Respetemos La Naturaleza (2)

This site is also known as 'the leaking pipe'. 9.6km northwest of Vilaflor there are two signs near some old buildings on the right, where the road bends to the left and crosses a barranco (dry gully). One sign states 'Respetemos La Naturaleza' and the other 'Peligro de Incendio'.

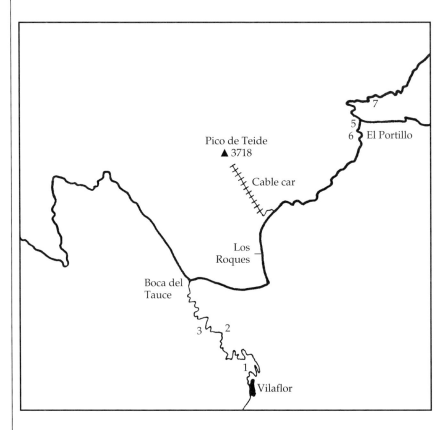

On the opposite side of the road a small indistinct track leads downhill to a water pipe which has a hole in it plugged with a piece of wood. When there is water leaking from this point the area is usually full of birds coming to drink. Look out for Great Spotted Woodpecker, Chiffchaff, Tenerife Goldcrest, Blue Tit, Blue Chaffinch and Canary. The occasional Raven passes overhead.

Las Lajas (3)

This is the best site for Blue Chaffinch. It is a very popular picnic area for locals and tourists alike so an early visit is advisable. 10.8km from Vilaflor and 1.2km from the previous site, it is possible to see the picnic tables, barbecues and the bar/restaurant on the left. Turn in at the main entrance signposted Zona Recreativa Las Lajas. Beware of the speed bumps and park where convenient. The Blue Chaffinch is often heard before being seen and sounds rather like a slurred Chaffinch. In order to see this species either walk towards the calls and find them in the trees or sit and wait for them to come down into the picnic area. They can often be seen around the water taps, as can Great Spotted Woodpeckers, especially during the hotter summer months. Other species which occur here include Kestrel, Turtle Dove (summer), Rock Dove, Plain Swift, Berthelot's Pipit, Chiffchaff, Tenerife Goldcrest, Blue Tit, Raven and Canary.

Tenerife
Goldcrest

Zona Recreativa Chio (4)

This is just off the main road between Chio and Boca del Tauce, 16.2km from Chio and 12.4km from Boca del Tauce, and is signposted, although the sign is rather obscure. It is a picnic area with barbecues and water taps but is a lot more open than Las Lajas. The species occurring here are very similar. This is a good site for Blue Chaffinch but some of the other species such as Great Spotted Woodpecker, Tenerife Goldcrest and Canary are less common here than at Las Lajas.

El Portillo (5)

The bar/restaurant here used to keep Blue Chaffinches in a small aviary. Thankfully, this is no longer the case. Blue Chaffinches do occur in nearby trees.

Las Cañadas Visitors Centre (6)

Just 200m towards Mount Teide from El Portillo, the gardens around the national park visitors centre and the two or three very small drinking pools can attract large numbers of Canaries and a few Blue Chaffinches.

Galeria Pino de la Cruz (7)

Situated between El Portillo and La Orotava. on the C821, 21.3km from La Oratava is another picnic area, albeit a very small one, which attracts Blue Chaffinches.

La Esperanza to El Portillo (8)

Any roadside stop after the 15km post on the C824 road from La Esperanza to El Portillo could produce Blue Chaffinch but they are a lot harder to find in the dense forest than at the various picnic areas. The lookout (mirador) near El Diablillo is one of the better places to look. This area is also very good for Tenerife Goldcrest as well as the more common pine forest species.

Mirador de la Cumbres (9)

Yet another site for Blue Chaffinch, just off the C824 between the turning to Arafo and the observatory at Izaña. This is also a superb viewpoint.

LA GOMERA

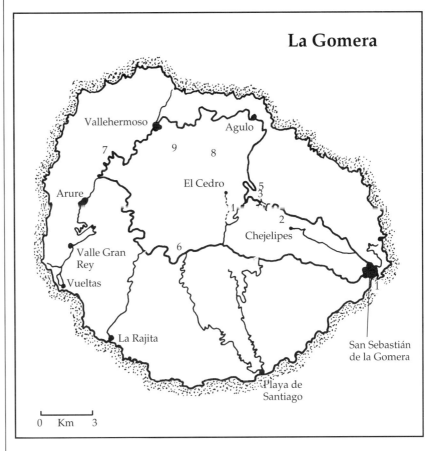

This island has the following specialities – Bolle's Pigeon, Laurel Pigeon, Plain Swift, Berthelot's Pipit, Tenerife Goldcrest and Canary. The two pigeons are usually much easier to see on this island than on Tenerife, especially the rarer Laurel Pigeon. It is the ferry crossing from Tenerife to La Gomera that most birdwatchers use in their search for seabirds.

One species not mentioned in any of the following Gomeran sites is Rock Sparrow. This species breeds on the island but is nomadic and hence does not have a regular site. It is always worth checking birds perched on overhead wires, particularly along the roads out of San Sebastian, as this is quite an elusive species in the Canary Islands.

The Gomera Ferry

Strategy   At the time of writing there are two ferries operating on the route between Los Cristianos (Tenerife) and San Sebastian de La Gomera. The times of sailing can vary with the season, so check the latest schedule on arrival in the Canary Islands. The approximate return fare for a foot-passenger is 3,000-4,200 pesetas (about £20) depending upon the time of the crossing. The crossing usually takes about one hour and twenty minutes, but this is dependant on weather conditions as it can be very windy and at times quite rough. The ferry used by the majority of

Bulwer's Petrel

P.J.

birdwatchers is the Benchijigua 2 run by the Ferry Gomera company. It is much larger than its competitor run by Transmediterranea. If travelling to Gomera in the morning and returning in the evening, then 'port out, starboard home' should be adhered to as this usually gives the maximum protection from the wind. It also seems preferable to be as near to the bow as possible as the extra height improves the viewing area.

Birds   The specialities to look for on this crossing are Bulwer's Petrel, Little Shearwater and Madeiran Storm-petrel. Small numbers of Bulwer's Petrel can be seen from early May through to early September. Little Shearwater is a winter breeder but a few can still be seen on most crossings during the summer. Madeiran Storm-petrel is a very rare winter breeder occasionally encountered in May and August. Cory's Shearwater is an abundant summer visitor and a few sometimes remain throughout the winter. British Storm-petrel is a scarce summer breeder usually seen between May and September. Any small white-rumped storm-petrels seen during the summer months are much more likely to be British Storm-petrel than Madeiran. Manx Shearwater is normally only seen on passage in spring and autumn. Gannets are seen in winter.

Rarities which have been seen from the ferry include Great, Sooty and Mediterranean Shearwaters, soft-plumaged petrel sp., Wilson's and White-faced Storm-petrels, Red-billed Tropicbird, Long-tailed, Great and Arctic Skuas, and Roseate and Sooty Terns.

San Sebastián de La Gomera (1)

Strategy Check the harbour area which is usually a summer site for one or two pairs of Common Terns. Scan the beach which sometimes attracts a few waders. Walk or drive along the seafront until the road turns right and heads inland, where there is a barranco that is normally dry. There are often a few pools in the culverts on the inland side of the beach which are always worth a quick look. Probably the best place for seawatching around San Sebastian is the end of the harbour wall.

The regular species around the town include Cory's Shearwater (offshore in summer), Kestrel, Yellow-legged Gull, Turtle Dove (summer), Plain Swift, Hoopoe, Berthelot's Pipit, Grey Wagtail, Spectacled Warbler, Blackcap, Chiffchaff, Blue Tit, Spanish Sparrow and Linnet.

The harbour and barranco have attracted rarer species such as Ring-necked Duck, Osprey, Barbary Falcon, White-rumped Sandpiper, Collared Pratincole, Laughing Gull, Gull-billed Tern and Pallid Swift.

Chejelipes Reservoir (2)

Location On leaving San Sebastián on the road to Hermigua and Vallehermoso there is a turning to Chejelipes about 0.5km after the left turn to Playa de Santiago and Valle Gran Rey. Take this turning and after passing the Dorada depot on the right, follow the road along the barranco. 5.6km after the Dorada depot is Chejelipes, from where it is possible to view the reservoir.

Birds The marshy area below the dam wall is one of the few breeding sites for Moorhen on La Gomera. This area does not normally hold much else of interest, but it is worth a quick look if time allows. White-rumped Swift has been recorded here.

Bar La Carbonera (3)

This is a good site for Bolle's and Laurel Pigeons.

Location From San Sebastian take the road to Hermigua and Vallehermoso. After about 7km the road deteriorates, and passes through three short tunnels. It then improves again before reaching a fourth, larger tunnel. Four hundred metres after this tunnel is the bar La Carbonera on the right. Park here and view the forested slopes opposite.

Birds Both Bolle's Pigeon and Laurel Pigeon can be seen from the patio or car park. Other species that can be seen from here are Sparrowhawk, Buzzard, Kestrel, Rock Dove, Turtle Dove (summer), Plain Swift, Berthelot's Pipit, Blackcap, Sardinian Warbler, Chiffchaff, Tenerife Goldcrest, Blue Tit, Robin, Blackbird, Raven, Chaffinch, Canary and Linnet.

Laurel Pigeon

Monte El Cedro (4)

This area is very good for Laurel Pigeon, much better than any of the Tenerife sites, and is good for Bolle's Pigeon as well.

Location

From the bar La Carbonera continue on for 1km and take the left turn to Monte El Cedro. Initially the road passes through mixed vegetation, interspersed with a few small fields, before starting to climb into purer laurel forest. It is worth stopping along this road at any of the various vantage points but be careful where and how you park as the road is used by both heavy vehicles and coaches.

Strategy

One of the good vantage points, with space for a couple of vehicles, is on the right 4.4km from the start of the road. After another 1km a walking track to Bosque de Tejos leads off to the left. There is room for a couple of cars to park here. Walk along the track to the bottom of the hill, taking care as it is quite steep and slippery due to loose rocks, to a green rain gauge, and look from here for passing pigeons. Return to your car and continue on the road 0.6km to the right-hand turning to El Cedro, a picturesque forested valley with a few houses. The road down into the valley is dirt, but is driveable except when very wet. It then gets very slippery, though it can still be managed with care. Along this road ignore the left turn after 1.5km to 'Barranco y Ermita' and continue another 0.5km to the entrance to the picnic area on the right. Park here and view the valley from the centre of the picnic area. This area is not as good as the vantage points along the main road, but you can still see both pigeons from here. From the top of the El Cedro road to the junction with the San Sebastián to Valle Gran Rey road is only 1.4km.

Birds

The species that can be seen in this area are the same as those for Bar La Carbonera.

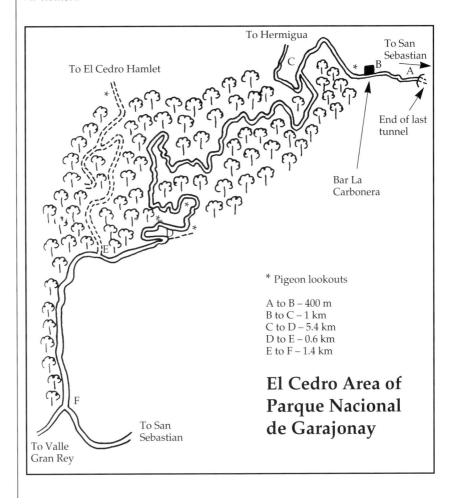

To Hermigua

To San Sebastian

To El Cedro Hamlet

C

B

A

End of last tunnel

Bar La Carbonera

* Pigeon lookouts

A to B – 400 m
B to C – 1 km
C to D – 5.4 km
D to E – 0.6 km
E to F – 1.4 km

El Cedro Area of Parque Nacional de Garajonay

F

To San Sebastian

To Valle Gran Rey

Presa de Los Tilos (5)

This small reservoir can be seen to the right of the road 1.4km after the turning to Monte El Cedro on the road to Hermigua. After another 400m there is a track which leads down to the dam. Be very careful stopping or parking on this section of road as it is windy and no one expects people to be stopped on the side of the road. Usually there is very little birdlife here as the barranco is steep-sided, but it is worth a quick look as at is one of the few areas of freshwater on the island.

Parque Nacional de Garajonay (6)

Situated in the centre of La Gomera, this national park contains a large area of laurel forest and holds an important percentage of the island's Bolle's and Laurel Pigeons. Most of the park is difficult to view except along the road to Monte El Cedro but there are a few other vantage points, such as the lookout on the San Sebastián to Valle Gran Rey road, 3.1km from the right turn to El Cedro and Hermigua, and about 100m after the turning on the left to Alajero. From here you are looking over the top of the El Cedro valley, and both pigeon species can occasionally be seen.

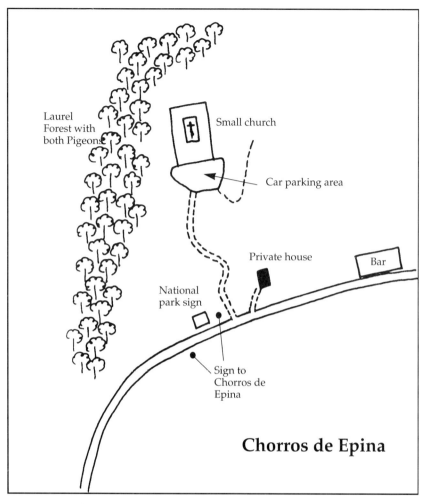

Laurel Forest with both Pigeons

Small church

Car parking area

Private house

Bar

National park sign

Sign to Chorros de Epina

Chorros de Epina

Chorros de Epina (7)

200m before the turning to Alojera (not Alajero) on the road between Valle Gran Rey and Vallehermosa is a bar called Chorros de Epina. Just before this on the left is a small track signposted to Chorros de Epina. Sometimes this track has a chain across its entrance, but fortunately not often. Drive along this track and park in the small car park in front of the square and the chapel. Look towards the ridge for the two endemic pigeons. Both species can usually be seen with ease at this site, but activity seems to be at its peak early in the morning. All the usual woodland/forest species can be seen within this area. The Chorros, a natural spring, can be seen by taking the walking track which leads off the corner of the little car park. The spring attracts birds to drink, and the local race of Chaffinch is particularly common here. This site is 17.5km from the El Cedro junction on the San Sebastián to Valle Gran Rey road.

Juego de Bolas (8)

This is the visitors centre for the national park and can be reached either from the road between Agulo and Vallehermosa or from near Laguna Grande on the San Sebastián to Valle Gran Rey road. This is the place for anyone requiring information on the park, but for birdwatchers another reason to come here is to look at the reservoir. Take the road that runs alongside the Bar Teraza and after about 1.7km the tarmac finishes and the road becomes a dirt track. The reservoir can be seen from here but closer views can be obtained by driving further along the dirt track. This is very similar to Presa de Los Tilos, being situated in a steep-sided barranco and is not all that attractive to birds.

Presa de la Encantadora (9)

In the centre of Vallehermoso take the turning to Triana. This is a fairly narrow road so be careful of oncoming vehicles, especially on blind corners. After 2.1km there is a small turning on the left which leads down to a reservoir. This is another Gomeran breeding site for Moorhen and as it is more open than some of the other reservoirs it is far more attractive to birds. This site is a little off the beaten track and only people spending more than a day on La Gomera should consider a visit to this reservoir.

LA PALMA

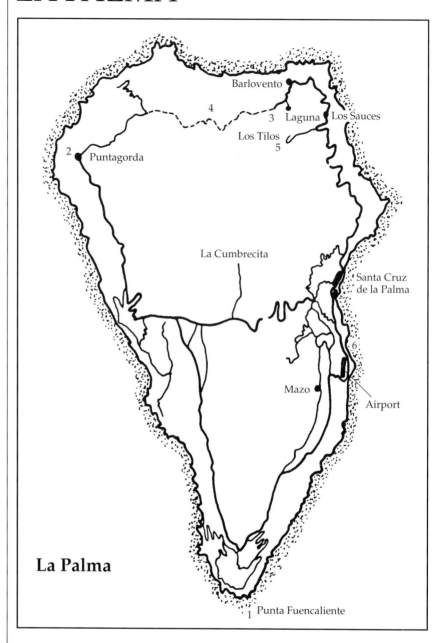

La Palma

This is probably the best island in the group to find the two endemic pigeons, and is the only island with a population of Choughs. Other specialities are Plain Swift, Berthelot's Pipit, Tenerife Goldcrest and Canary (very common). La Palma is definitely under-watched and is hardly ever visited by birdwatching tourists. The sites mentioned here are the main birdwatching areas on the island but the potential for discovering other sites still exist.

Punta Fuencaliente

This is the southern tip of the island and is good for seawatching.

There are also some salt-pans here which must have potential for attracting waders at the right times of year, though they have been seldom visited by birdwatchers. It was on the beach at nearby Playa Nueva in November 1991 that a Waterthrush sp. was recorded, proving the potential for vagrants on this island.

Location   Coming from Santa Cruz de La Palma turn left (south) at the Cepsa petrol station in Fuencaliente and follow the road out to the lighthouse and saltworks.

Birds   As so few birdwatchers have visited this site no bird list exists as yet.

Puntagorda Reservoir

This is a concrete-banked reservoir, but as it is one of the few areas of fresh water on La Palma, regularly attracts birds.

Canary

Location   Coming from the south, take the second of the two left turns opposite the Colmegran S.L. building on the outskirts of Puntagorda. Continue along this road to a large pine tree. Turn left here and then turn right by the Bar Paradilla El Pino. Follow this road until the reservoir becomes clearly visible.

Birds   As so few birdwatchers have visited this site no bird list exists as yet.

Laguna de Barlovento

This is one of the most well-known sites on the island. It is a man-made, concrete-banked reservoir which is attractive to birds, is regularly used by gulls for bathing and attracts ducks in the winter. The surrounding habitat attracts a good selection of the island's birds, including the two endemic pigeons. The area also holds races of Blue Tit and Chaffinch which are endemic to La Palma.

Location   The reservoir is situated in the northeast of the island, west of Los Sauces, and south of Barlovento. Take the main road south out of Barlovento and take the left turn signposted to Laguna de Barlovento. The reservoir can be viewed from the road bordering the eastern and southern sides. The wooded areas can be found by following the tarmac road to its end and continuing to the end of the major dirt track.

Birds   Other species in the area include Sparrowhawk, Buzzard, Kestrel, Quail, Woodcock, Yellow-legged Gull, Rock Dove, Turtle Dove (summer), Plain Swift, Long-eared Owl, Berthelot's Pipit, Grey Wagtail, Robin, Blackbird, Sardinian Warbler, Blackcap, Chiffchaff, Tenerife Goldcrest, Chough, Raven, Canary and Linnet.

The Northern Barrancos

This is an area of many wooded barrancos and is perhaps one of the most scenic regions in the whole of the Canary Islands. Both Bolle's and Laurel Pigeons can be seen very easily as they are even more common here than on La Gomera.

Location   This area lies between Barlovento in the northeast and Puntagorda in the northwest. At the time of writing the road joining these two towns is still no more than a good dirt track for most of the way, although there is more tarmac than shown on the majority of road maps. However, the long term plan is to tarmac the road completely and there are road works affecting access. It would be wise to seek advice in Barlovento or Puntagorda about the road's condition before attempting to cross or enter this area.

Birds   The birds found here are the same as in the woodland areas previously mentioned for Laguna de Barlovento.

Los Tilos

This is another wooded barranco and is an excellent site for both Bolle's and Laurel Pigeons as well as the other typical woodland species. This is arguably the best site on the Canary Islands to see these two species.

Location   If coming from Santa Cruz de La Palma, turn left into the barranco, about 2km southwest of Los Sauces, at a signpost to Los Tilos.

The two species of pigeon can easily be seen anywhere along this road. At the end of the road there is a bar where you can enjoy good food and/or a drink while glimpsing the pigeons through holes in the canopy. This is off the usual birdwatching circuit, but those with time to spare will not be disappointed by a visit to this marvellous spot.

The Airport Pools

Just to the north of the airport car park, and visible from the airport road when coming from Santa Cruz de La Palma, these salt water pools provide a roosting area for waders and are well worth visiting during a stay on the island.

Spectacled
Warbler

EL HIERRO

This island is the least visited by birdwatchers and is unlikely to be visited by those who come to the islands for just one or two weeks. The specialities are Bolle's Pigeon, Plain Swift, Berthelot's Pipit, Tenerife Goldcrest and Canary, and breeding seabirds including Cory's Shearwater, Little Shearwater, Bulwer's Petrel and British Storm-petrel. The following are some of the better sites, which are described only briefly. Further exploration will undoubtedly discover other sites of note.

Valverde to Frontera

This road passes along a ridge, the northern slopes of which are, in places, covered in laurel forest. It is here that one can see Bolle's Pigeon, when the area is not completely shrouded in mist. Other species that can be seen along this road include Sparrowhawk, Buzzard, Kestrel, Woodcock, Rock Dove, Turtle Dove (summer), Plain Swift, Barn Owl, Long-eared Owl, Berthelot's Pipit, Sardinian Warbler, Blackcap, Chiffchaff, Tenerife Goldcrest, Blue Tit, Robin, Blackbird, Raven, Corn Bunting, Chaffinch, Canary, and Spanish Sparrow.

Charco Azul

These man-made lagoons are potentially good for waders and waterfowl at the right time of year. To the west of Frontera, turn north in Los Llanillos and follow the road to its end to view the lagoons.

Charco Mansa

On the northern tip of the island is another good lagoon with potential for migrant waders. Take the road north from Echedo which finishes at Charco Mansa.

Restinga

This is the southernmost tip of the island and is an excellent site for seawatching. The rocky coast attracts waders such as Turnstone, and the adjacent rubbish dump holds large numbers of gulls and is good for Raven.

Faro de Orchilla

This is an alternative seawatching spot to Restinga, but it is far more inhospitable, with little or no shelter. The road here from El Pinar, although not in very good condition, passes through pine forest where Canary is common and there are also reasonable numbers of the Tenerife Goldcrest.

Spanish Sparrow

GRAN CANARIA

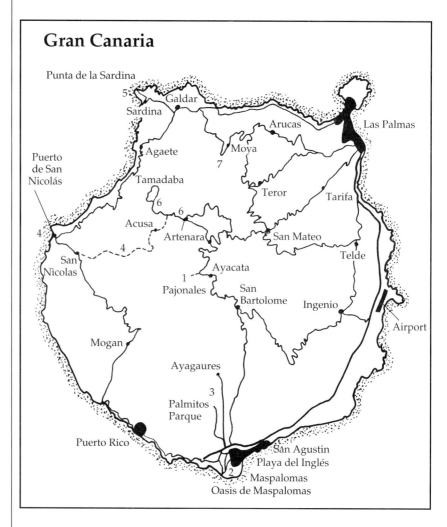

Gran Canaria

Gran Canaria is not regularly visited by many birdwatchers since it holds few of the endemic species. However, the pine forests high in the centre of the island support the second race of Blue Chaffinch, although it is by no means easy to see. Other endemics which are to be found on the island are Plain Swift, Berthelot's Pipit and Canary. Rock Sparrow occurs around some of the higher villages and gorges, and Lesser Short-toed Lark and Trumpeter Finch occur in the drier southern part of the island.

Pajonales (1)

The Pajonales Nature Reserve is the only reliable site for Blue Chaffinch, the island's most important bird. The reserve is on a plateaux averaging about 1,000m and contains extensive tracts of native pine forest. The high central area of Gran Canaria, of which the Pajonales area forms an extension, holds a montane flora with many endemic shrubs and herbs.

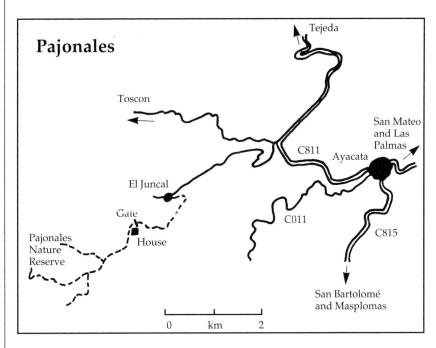

Location

The forest lies to the west of the central massif of the island, and to the southwest of the mountain town of Tejeda. The easiest access to the area is via the village of El Juncal. El Juncal is not served by bus, but can be reached by taxi from Tejeda or Ayacata. Both of these towns can be reached by bus. However, a hire car is much the best option. Approaching from the tourist resorts in the south, drive to San Bartholomí and take the C815 to Ayacata. Proceed towards Tejeda on the C811, and after about 3km turn left towards El Juncal and Toscon, then go left again to El Juncal. Drive into the village, continue down to the left and across the valley. The road is then gravel, but it is possible to drive to the entrance of the Pajonales Reserve, which is marked by a sign at the ICONA forest house.

Strategy

This site is best visited in spring (April, May or possibly June), when Blue Chaffinch is most likely to be in song. It is found mainly in areas of mature Canary Island Pine with an understory of the white-flowered shrub *Chamaecytisus proliferus*, locally known as Escobon.

If it is singing, Blue Chaffinch can be located relatively easily, since the song is very like the song of the Chaffinch, the endemic race of which does not seem to occur in this forest. Early mornings or evenings are likely to be the best times to hear it. Even in this forest, Blue Chaffinch appears to be at low density, and the bird is somewhat retiring, so it is difficult to locate when not singing. Concentrate on the areas with a good understory, such as the area up to the left, just beyond the ICONA house.

Birds

In addition to Blue Chaffinch, the pine forest holds endemic races of Great Spotted Woodpecker, Blue Tit, and Robin, and the inevitable Berthelot's Pipit. Plain Swifts occur overhead, and Buzzard and Raven

Berthelot's Pipit

are common. Other species include Sardinian Warbler, Rock Sparrow, and Canary.

Maspalomas (2 and 3)

The area around Maspalomas encompasses a wide variety of habitats, including a reed-fringed tidal lagoon, sand dunes, ornamental gardens, arid plains and valleys. This combination ensures that a good selection of the commoner species occurring on the island can be seen in the vicinity. There are two main birdwatching areas near Maspalomas; the Oasis de Maspalomas to the southwest of the town, and the valleys and hills with fairly natural semi-desert habitats to the north, around a 'Bird Park' at Los Palmitos. The two sites are dealt with separately. The area around Maspalomas is also the only known site in the Canary Islands for Tree Sparrow, although the exact location of breeding pairs varies from year to year.

Oasis de Maspalomas

The Oasis de Maspalomas centres around a tidal lagoon called La Charca. In the earlier part of the 20th century, the lagoon was extensive, and supported breeding Marbled Ducks. Much of it has now been reclaimed, but it is still one of the more interesting coastal areas on the island, and a variety of waders is usually present.

Location   The Oasis de Maspalomas is at the extreme southern tip of Gran Canaria, a few kilometres southwest of the main resort of Maspalomas/ Playa del Ingles.

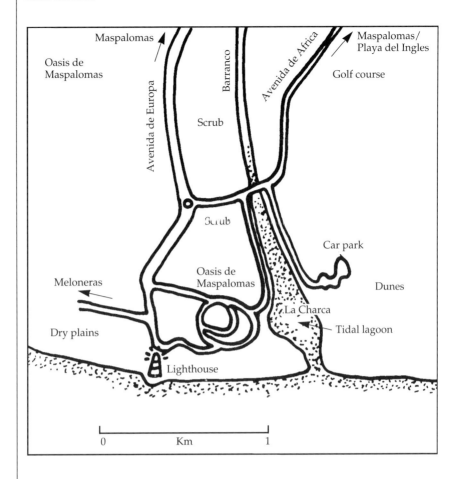

Strategy   Early morning is the best time to visit, before the birds have been disturbed by tourists. Later on, the waders may be rather more distant, or may disperse to other areas nearby.

After checking the lagoon, it is also worth looking in the gardens around the Oasis where a range of common species can be found. The open area near the lighthouse and the edge of the dunes can also be worth checking.

Birds   Waders which are often found around the lagoon include Sanderling, Dunlin, Kentish and Ringed Plovers, and Whimbrel, while other species such as Curlew Sandpiper and Little Stint occur on passage. Moorhen breeds in the area. The gardens around the Oasis hold breeding Starling, a species which is very local in the Canary Islands (it is also found in a few places in Tenerife), together with Hoopoe, Sardinian Warbler, Spanish Sparrow, and the local races of Chiffchaff and Blackbird. Great Grey Shrike occurs in the scrubby areas just north of the oasis, and in the dunes to the east of the lagoon, where Berthelot's Pipit is also found. The open areas near the lighthouse hold Lesser Short-toed Larks. Both Pallid and Plain Swifts can be seen over the lagoon, while hirundines occur on passage. In spring, passerine migrants sometimes occur in the area.

Los Palmitos

The bird park at Los Palmitos will not be to everyone's taste, but it does hold an extraordinary selection of the world's birds, including a particularly comprehensive collection of parrots. Some birds are free-flying, including lovebirds and parakeets. The valley of Los Palmitos (both above and below the bird park) contains relatively natural stands of Euphorbia scrub typical of the dryer parts of the Canary Islands, and a range of typical birds can be seen.

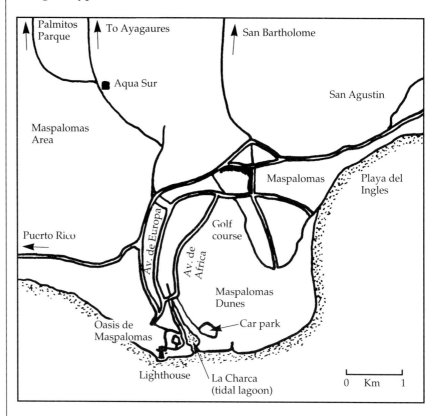

Location

To reach Los Palmitos, follow signs from Maspalomas towards the zoo, which later become signs to Palmitos Parque. En route to the bird park, it is worth stopping from time to time to check the semi-desert scrub and rocky areas. Trumpeter Finches occur in the area, together with a range of other species. Buses run from Maspalomas to the bird park, and this is one way to get out into good habitat if staying at the resort without private transport.

Nearby, the Ayagaures valley is also of some interest. Proceed from Maspalomas as before, but turn right just after the Aqua Sur pleasure complex. It is possible to drive all the way to Ayagaures village, which is just below the dam of the reservoir. However, be very careful on this road as it is narrow and winding. The valley has stands of tall, bamboo-like grasses and a reservoir.

Birds

Commoner species to be seen in this area include Kestrel, Turtle

Dove, Plain Swift, Berthelotís Pipit, and the local races of Blue Tit and
Chiffchaff. Common Waxbills can also be seen here, both around the
park and in the Ayagaures valley. This introduced species is now an
established resident in the area.

The San Nicolás area (4)

Just south of the small resort of La Aldea at Puerto de San Nicolás,
there is a small, shallow lagoon with tamarisk scrub which attracts
waders and passerines. The gorge inland from San Nicolás runs deep
into the mountainous heart of the island, is scenically very attractive,
and is a good site for Rock Sparrow. In addition to cliffs, this valley
holds three reservoirs.

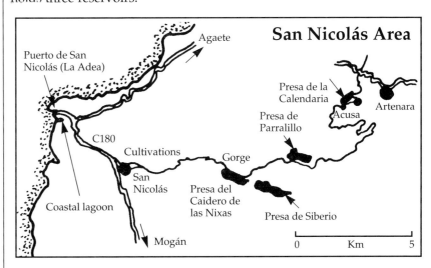

Location

La Aldea is a relatively remote resort at the western edge of the
island. All the roads which lead to it are winding, and the area takes a
considerable amount of time reach from any direction. It is perhaps best
approached from Agaete in the north. San Nicolás lies inland southeast
of La Aldea along the C810.

Strategy

From the main west coast road take the turning to Puerto de San
Nicolás and bear left to the car park by the Bar El Chozo. Continue on
foot beyond the car park and the lagoon is a short walk away along the
beach. It is possible to walk round the lagoon, checking the tamarisks
on either side. From here drive southeast on the C810, then enter San
Nicolás and head through the town towards Artenara. East of San
Nicolás the road passes through cultivated areas for a few kilometres,
then enters the gorge of the Barranco de la Aldea. The road becomes
very narrow and is also very steep in places, requiring a certain amount
of daring to complete the journey to Artenara. There are suitable
parking places from time to time, including a lookout over the first
reservoir (Presa del Caidero de las Niñas) and near the dam of the
third reservoir (Presa de El Parralillo). Check the cliffs as well as the
edges of the reservoirs.

Birds The lagoon at Puerto de San Nicolás attracts migrant waders such as Common Sandpiper, and the tamarisks and other vegetation in the area around the resort hold birds such as Turtle Dove, Sardinian Warbler, Blackcap, Chiffchaff and Canary. The area may be attractive to migrant passerines under the right conditions. Rock Sparrows occur in the gorges above San Nicolás and at the village of Acusa (e.g. around the church) while Ravens and Buzzards circle overhead. Trumpeter Finches occur lower down around San Nicolás and at the coast in the vicinity of Puerto de San Nicolás. Grey Heron and Moorhen can be seen on the reservoirs, while Grey Wagtails occur throughout the gorge and along the barranco to the sea.

Punta de la Sardina (5)

The northwest tip of the island is a good place for seawatching.

Location The point is north of Sardina. Follow signs to the lighthouse (Faro), and view from the cliff top next to it.

Cory's Shearwaters can be seen in large numbers fairly close to shore between March and September. Other seabirds might also be found by patient watchers, although the potential of the site is largely unexplored so far.

Pinar de Tamadaba and Artenara (6)

The pine forest at Tamadaba is rather easier to access than the Pajonales forest, and is on the main tourist route. In addition to the pine forest (mainly planted) there are areas of open montane scrub. The area holds the commoner forest birds as well as small numbers of Blue Chaffinch.

Location Tamadaba forest is reached via the mountain town of Artenara, which is to the northwest of the central massif. From Artenara, follow signs to the Tamadaba forest, and follow the one way road to the car park at the far end. It is possible to reach Artenara by bus from Las Palmas.

Strategy Artenara, as well as being a good place to stop for refreshments, is also worth checking for Rock Sparrow (especially around the church). Moving on to Tamadaba, the best place to park is in the car park at the far end of the one way loop road. There is a small flat area of pine forest here where it is relatively easy to birdwatch, and there is also a track leading from the car park area down into an area of pine plantation. However, if time permits, the best option is probably to walk round the one way loop (about 5km) checking the forest on either side. Alternatively, just walk part of the way and return to the car.

Blue Chaffinch occurs in the forest but appears to be very scarce and not at all easy to locate. The local race of Great Spotted Woodpecker is, however, both common and surprisingly confiding. Other birds which can be seen in the forest include Buzzard, Robin, Blue Tit, Chiffchaff, and Canary. Rock Sparrows are common around Artenara.

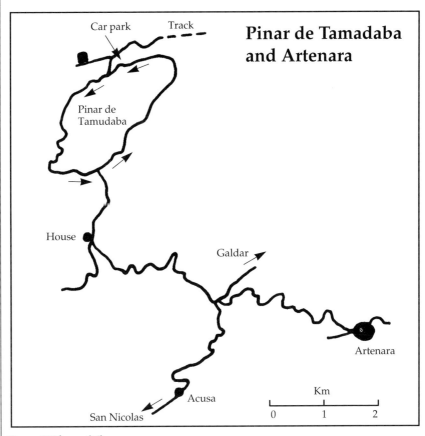

Los Tilos (7)

This is the last remaining area of laurel forest on Gran Canaria and many of the trees, shrubs and flowers growing in the wood are endemic. Although only small, and in a degraded condition, it holds a good range of commoner birds. Sadly, the endemic pigeons have long since disappeared. Although there are no special birds to be seen here, the site is worth a detour if travelling along the northern part of the island.

Location Los Tilos is on the Barranco de Moya, which is just to the west of the village of Moya. From Moya, follow the GC150 towards Guia and Galdar and after about 2km turn left to the Laurisilva (Laurel forest). There is a good parking area 800m from the junction.

Strategy Walk back along the road from the car park towards the junction, and view the birds in the wood to either side. The walk up the valley through more open country is also pleasant.

Birds The wood supports Canary, the distinctive endemic races of Chaffinch and Blue Tit, together with Turtle Dove, Blackbird, Robin, Sardinian Warbler, Blackcap, and Chiffchaff. Other species which may be seen in the area include the endemic races of Buzzard and Grey Wagtail. Plain Swift and Berthelot's Pipit can also be seen.

Other Wildlife Butterflies include the Canary Speckled Wood.

FUERTEVENTURA

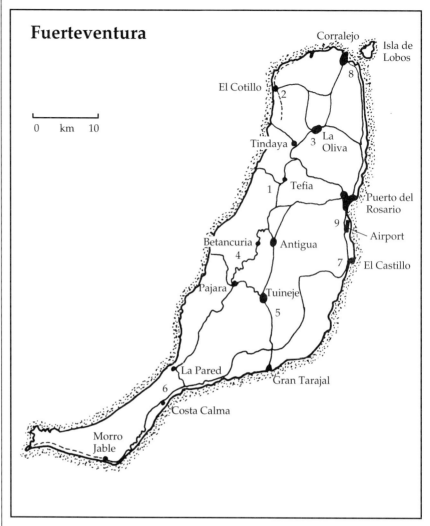

Fuerteventura's most famous bird is the Canary Islands Chat, a species which is entirely restricted to this island. Although local, any visiting birdwatcher should be able to find this species provided that the right areas are investigated. The island is also home to a number of desert birds, the most important of which is the Houbara Bustard. This species is probably commoner in Fuerteventura than in any other comparable area in the world. Other desert species which are fairly common include Cream-coloured Courser, Black-bellied Sandgrouse and Trumpeter Finch. Egyptian Vultures are still fairly numerous. Of the other endemic birds, only Berthelot's Pipit is common. Plain Swift occurs, but Pallid Swift is much more numerous. Lesser Short-toed Lark is abundant.

Los Molinos (1)

The reservoir at Los Molinos is the largest body of freshwater in the eastern Canary Islands and the only one which never dries up. A trip

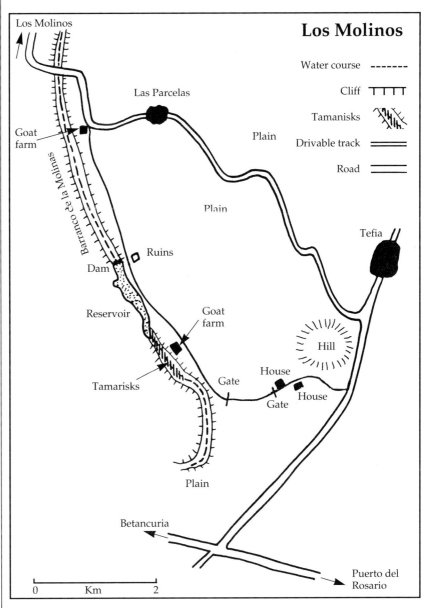

to this site is a must for all birdwatchers visiting Fuerteventura. It is a good site for waders, and has produced a number of rarities. A good range of resident species can also be seen around the reservoir.

Location   The reservoir is near the west coast of the island to the west of the village of Tefía. There is no bus service to this area, and it is best visited by car, although it should be possible to hitch to Tefía.

Strategy   The site is worth a visit at any time of the year since a number of the island's specialities occur in the area. However, spring and autumn are best since a wide selection of waders is often present, and the tamarisks and other vegetation attract passerine migrants.

Resident species can be looked for on the plains to the west of the road between the main Puerto del Rosario-Betancuria road and Tefía. Early morning and late evening are the best times. The permanently wet barranco below the reservoir dam is also worth a look for both residents and migrant species.

Access to the reservoir itself is via one or other of two fairly rough but quite driveable tracks. The track to the dam starts from the small settlement on the Tefía to Los Molinos road now known as Las Parcelas, but previously called Colonia Garcia Escamez (and marked as such on some maps). This track is on a sharp right hand bend by a goat farm 5.5 km from the Tefía end of the road. The dam is 2.5 km south along the track. Park by the dam and walk along the ridge towards the inlet end. It is best to stay above the reservoir since the birds are easily flushed due to the narrowness of the water. The small inlet not far up from the dam on the opposite shore often holds a variety of waders and can be checked with a telescope. Then proceed towards the tamarisks at the inlet end, dropping down to follow the inlet stream and to check the tamarisks for passerines. The stream can be followed for several hundred yards to the point where it vanishes underground (except immediately after heavy rain).

The second track leads to the other end of the reservoir, i.e. the inlet end. If approaching from the south, turn off the main Puerto del Rosario-Betancuria road towards Tefía, and take the track on the left after 2.5 km. If approaching from the north, pass through Tefía, continue past the turn to Los Molinos (signed to Las Parcelas) and the track is another 0.9 km on the right. The track passes two houses and then goes through two gates. Please close these gates securely behind you. It is possible to park on the left just after the second gate. From here, drop down into the barranco and follow it to the right. Walk down beside the stream to the reservoir. Pass the goat farm up on the ridge to the right and then keep to the right of the stream. Then follow the goat path round the edge of the tamarisks, and up onto the hillside above the reservoir. Alternatively, drive on past the goat farm and park nearer to the reservoir.

Birds The barranco between the coast and the reservoir supports several pairs of Canary Island Chats, which are perhaps best looked for in the area a few hundred metres downstream from the dam, where waders also occur. Plain Swift apparently nests in the barranco's cliffs.

The plains hold small numbers of Houbara Bustard and Cream-coloured Courser, while Stone-curlew is fairly common. The area near the main Puerto del Rosario-Betancuria road is perhaps best for Houbara Bustards.

However, it is the reservoir itself which is the main centre of ornithological interest. This is the only site in the Canary Islands where there is a good chance of seeing Marbled Duck. If present, they usually keep to the flooded tamarisks and can be difficult to see. The reservoir supports good populations of Coot and Moorhen, both of which are very local in the Canary Islands. It is worth checking for Crested Coot, a species which has been reported here in the past. Little Egrets are usually present, and Night Herons are regular on passage. At least a few waders are always to be found, particularly at the southeast end and along the inlet stream. Little Ringed Plovers breed regularly, and Kentish Plovers breed occasionally. Species such as Snipe, Redshank, and Green and Common Sandpipers are present in winter. A wide range of waders occurs on passage, and good numbers of Black-winged Stilts have recently been recorded. Three species of crake have been reported.

The site has proved to be very attractive to migrant passerines in the past, and species which have been seen include Bee-eater, Nightingale, Redstart, Pied and Spotted Flycatchers, and a variety of warblers. Unfortunately, the tamarisks have recently died and the area is likely to be less attractive for migrants in the future. However, it is still worth checking the area for migrant passerines during February-May and August-October. Wintering passerines which have been recorded include Song Thrush and Chiffchaff (various races).

Rarities which have been recorded at this site include Little Bittern, American Wigeon, Blue-winged Teal, Scaup, Booted Eagle, Allen's Gallinule, Whiskered Tern, Red-rumped Swallow, Thrush Nightingale, Subalpine Warbler, and Woodchat Shrike.

El Cotillo (2)

The area around the small fishing village of El Cotillo contains several different habitats. To the south there is a dry, stony, coastal plain, backed by low hills. This is the closest area to Corralejo where Houbara Bustard can be reliably found. To the northeast is a partially vegetated lava field which stretches all the way to Corralejo. The coast itself is rocky but with shallow, sandy bays.

Location

El Cotillo is in the northwest of the island, 17km from La Oliva and within easy reach from the resort of Corralejo by car. There is a daily bus service from both Puerto del Rosario and Corralejo, but an overnight stay would be required to be in the area at dawn. There is no short-term accommodation in the village, but it is possible to camp among the dunes to the north of the village. This area is popular with local people, many of whom camp out here at weekends.

Strategy

The area is good for resident birds but, with the exception of the shore, is not very attractive to migrants. Waders are likely to be more numerous in winter and during the migration period. Consequently, there is little difference in the strategy to be adopted at different seasons.

Early morning should be spent on the plain to the south of the village since the species to be seen there are more visible at this time. A good track follows the coastal fringe of the plain, and it is possible to drive all the way to Barranco de Esquinzo. On arriving at El Cotillo, turn left onto the track just as you enter the village, and continue past walled cultivations and a few scattered buildings. After 3km stop and scan the plains inland for desert birds, using the car as a hide. There are driveable tracks off to the left, but it is probably just as profitable to stay on the main track, stopping to scan every few hundred yards. After an hour or so of this approach, it is worth getting out and walking towards the low hills which mark the landward edge of the plain.

The sandy bays and rocky shores to the north of the village are good for waders. It is worth going at least as far as the lighthouse. The lava field begins a little way inland from the sandy area, and may also be worth a look. Walk round the edge rather than trying to cross it since the lava field is very rough going and is covered in a never ending series of high walls. The area to the north of the turning to Lajares is perhaps as good as any. Inland, the extensive sandy plain near Lajares is also worth a look.

Seawatching is worthwhile (early morning or evening), and a good place to watch from is the promontory opposite Restaurante Bar Azzurro, about 1.5 km north of the village.

While there is a driveable but very rough track along the whole of the shore north of the lighthouse, beware of the soft sand on either side.

Birds   The plain to the south of the village is as good as anywhere on the island for Houbara Bustard, and it is not unusual to see half-a-dozen in an hour or two. Cream-coloured Coursers are also present, and Stone-curlews are common throughout the area. Egyptian Vultures may sometimes been seen overhead. Berthelot's Pipits and Lesser Short-toed Larks are abundant, and Spectacled Warblers occur in scrubby areas.

Cream-coloured
Courser

The shore to the north of El Cotillo usually holds a Little Egret or two, and waders such as Sanderling, Turnstone, Whimbrel, and Grey and Kentish Plovers are usually fairly common. Offshore, Cory's Shearwaters are numerous between March and September, especially in the evening, and rarer species might be seen with luck.

Although the lava field does hold a few pairs of Canary Islands Chats, this is not a good area to look for them. However, Barbary Partridge, Hoopoe, Spectacled Warbler, Great Grey Shrike, and Trumpeter Finch all occur in fair numbers.

The sandy plain to the south of Lajares also holds Houbara Bustards, but the shrubby hummocks make the species hard to find here. Cream-coloured Coursers also occur here, and a few Black-bellied Sandgrouse may be found.

La Oliva (3)

La Oliva is the main agricultural village in northern Fuerteventura. The four habitats in the vicinity are; cultivated areas around the village itself, mountains and valleys to the east and south, plains to the west, and lava fields to the north. A wide selection of the island's resident birds can be found here as well as many migrants.

Location   La Oliva is in the centre of northern Fuerteventura, 17km south of Corralejo. There is a convenient daily bus service from Corralejo, and there are also daily buses from Puerto del Rosario.

Strategy   The area around La Oliva holds most of the island's resident species, and is good at any time of year. However, the area really comes into its own during the spring, when crops and bushes attract good numbers of migrants.

Birds   The cultivations south of the village hold good numbers of many of the common species, including Hoopoe, Berthelot's Pipit, Spectacled Warbler, Great Grey Shrike, and Trumpeter Finch. Relatively uncommon birds which can also be found here include Goldfinch and Corn Bunting, while Quail sometimes nest in years following heavy winter rains, when crops are relatively lush. The eastern Canary Islands race of Blue Tit may also be seen in trees around the village. Cultivations north of the village (left of the Corralejo road and behind the houses) can also be good for migrants.

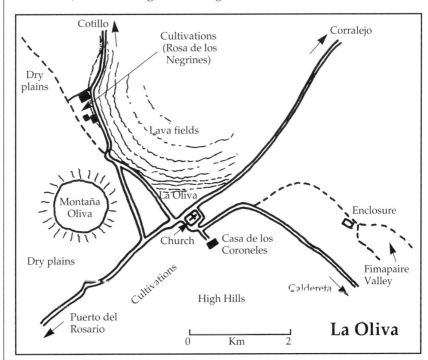

Houbara Bustard, Cream-coloured Courser and Stone-curlew occur on the plains and rolling hills to the south and west of Montaña Oliva, where Lesser Short-toed Larks are common. However, the best area for the bustard is northwest of the village, and is reached by following a driveable track to the left after the last house in the village along the Cotillo road (on a right-hand bend). This track passes an area of cultivations with scattered palms and fig trees on the right. This is the area known as La Rosa de los Negrines. Further on is a series of low hills which supports a high density of Houbara Bustards. In the evenings they often walk to the area of cultivations to feed, and this can be an excellent place to watch them. Try parking the vehicle beside the cultivations, in the area where there are low stone walls on either side. Then simply watch from the vehicle. The cultivations are also good for migrants in spring when the area is an 'oasis' sandwiched between the plains to the south and the lava fields to the north.

In winter the resident birds are joined by small numbers of northern migrants such as White Wagtail, Song Thrush, Blackcap, and Chiffchaff. Other species such as Lapwing, Redwing, and Black Redstart also occur from time to time. Rarer species have been recorded, including White Stork and Dotterel.

The whole area is attractive to migrants in spring, and to a lesser extent in autumn. The commoner species include Tree Pipit, Redstart, Whinchat, Willow and Wood Warblers, and Pied Flycatcher. Other species such as Night Heron, Bee-eater, Wryneck, and Nightingale are also fairly regular, and rarer species have included Red-rumped Swallow, Tawny Pipit (has wintered), Bonelli's Warbler, and Ortolan and Cretzschmar's Buntings.

To the east of the village, a series of deep valleys leading to the east coast provide the most northerly area to see the Canary Island Chat, and this is the nearest reliable site for anyone staying at Corralejo. Perhaps the easiest spot to see them is at the head of the Fimapaire valley in the area where the track diverges. From La Oliva take the road beside the church which heads towards the Casa de los Coroneles, and follow it round to the left towards Caldereta. One kilometre from the church, after a right-hand bend, turn left onto a firm track, and follow this for a further 2km, passing a fenced enclosure on the right. The track forks just beyond this enclosure, at a site used as an unofficial car dump. The chats are normally to be found in the immediate area, or towards the bottom of the slopes on either side.

This area, together with the old lava field in the valley bottom, is also good for Barbary Partridge, while the extensive and more recent lava fields to the north of La Oliva are also good for this species (and Trumpeter Finch).

Other wildlife   The introduced Barbary Ground Squirrel is common in this area. Butterflies found in the crop areas include specialities such as African Grass Blue, Greenish Black-tip, and Green-striped White.

Las Peñitas (4)

The mountains around Betancuria, the ancient historical capital of the island, provide a welcome contrast to the desolate plains which cover much of the island. South of Betancuria is the village of Vega de Río

Palmas, a relatively lush area with many palm trees and cultivations. The Barranco de las Peñitas runs through the valley and is dammed below the village to provide a virtually permanent lake nestling below steep rocky mountains. The shallow margins of the lake support a dense tamarisk thicket.

Location Vega de Río Palmas is on the Pájara-Betancuria road. Hire car is by far the easiest way to get there, although hitching should also be possible. The roads in the area are typical mountain roads.

Strategy The road to the lake is signposted to Vega de Río Palmas from the main road, and ends just above the lake. However, the land between the road and the lake is private, and there are 'no entry' signs at the two parking places. In any case, the barranco itself is attractive to migrants since there is normally a trickle of water in it, and there are bushes and trees providing cover. It is therefore better to park near the bridge and walk the 1.5km or so to the lake. Where the barranco reaches the lake, walk up the hill about 100m where a track then leads to the dam, giving views down to the lake. The tamarisk thicket is another good area for passerines. The area is sheltered and tends to be

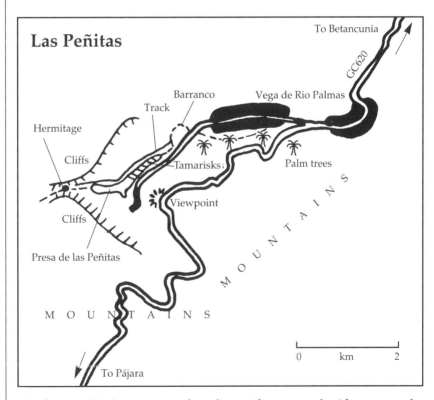

quite hot, so it is important to be adequately prepared with water and sun cream before setting out on this walk.

Birds The Canary Islands Chat is relatively common in the hills around Betancuria, and both the barranco and the area around the lake itself are good places to see it. Another good area is beside the road between Vega de Río Palmas and Pájara, particularly between Km27 and Km24. However, be careful where you park since the road is narrow and dangerous.

The cliffs which surround the lake attract raptors, and both Egyptian Vulture and the Canary Island race of the Buzzard should be seen. Barbary Falcons are also seen in the area from time to time. The local race of Barn Owl, which is difficult to see, has been reported from this area, and the rocky slopes provide what is perhaps the best area on the island for Barbary Partridge. The viewpoint along the road to P·jara, which is directly above the reservoir, provides a good vantage point from which to watch for raptors and may produce the partridge (on nearby slopes) as well.

Sardinian Warblers are common in the tamarisks around the lake and along the barranco, and the species also occurs in the gardens at Betancuria and P·jara. The Eastern Canary Islands race of the Blue Tit is fairly common in the area. This is one of the few areas in the island where Greenfinches are often seen.

The lake itself currently holds a large colony of Coots, and Little Ringed Plovers also breed. A visit in winter, spring or autumn usually produces a few waders, such as Snipe, Common and Green Sandpipers, Redshank or Greenshank. Other migrants/vagrants which have been recorded around the reservoir or in the barranco include Night Heron (regular), Squacco Heron, Little Egret (occasional), Tufted Duck, Pintail, Marsh Harrier, Bee-eater, and Ortolan Bunting.

The park in the centre of Pájara, as well as the trees around the church, are attractive to wintering and migrant passerines, and the area is well worth a quick look if you are passing through. Migrant and wintering passerines which have been recorded here include Wood Warbler, Sedge Warbler, Redwing and Siskin.

Other wildlife   Barbary Ground Squirrels are common in the area. The barranco between the bridge and the lake is the best site in the island for the spectacular Plain Tiger butterfly, and a good range of other butterflies can also be seen. Just below the dam, the tiny hermitage of Nuestra Señora de la Peña is of cultural interest and worth a short detour.

Catalina García (5)

A small lagoon has recently been formed by damming a shallow section of the Gran Tarajal Barranco. The margins are shallow and free from tamarisk growth, and are ideal for waders. Nearby habitats include cultivations, an extensive tamarisk thicket and the southern end of the dry central plain.

Birds   Catalina García is in the south of the island, about 3km south of Tuineje on the Gran Tarajal road. From Gran Tarajal drive just north of the 9km post, turn right onto the track beside a goat farm (the lagoon can be seen from here), and then turn right again, passing the lagoon on the left.

On the opposite side of the road is the cultivated area known as Rosa de Catalina García, which has a variety of trees and shrubs, with a tamarisk thicket in the barranco to the south. To the northeast of the lagoon there is a dry plain.

Strategy   Drive round the lake viewing from the car, and continue across the earth dam. At the far end of the dam the track ends, and there are good views of the marshy areas from here.

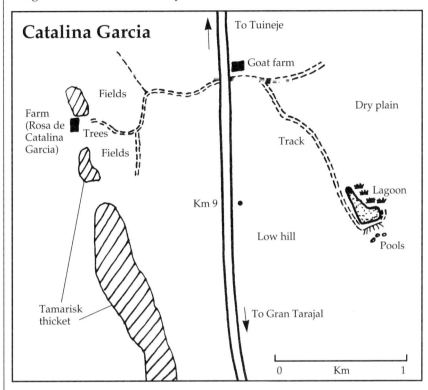

If time permits, the cultivated area on the other side of the Gran Tarajal-Tuineje road may also be worth a look, particularly during the migration period. The area should be viewed from the track. Do not enter the cultivations themselves without asking permission at the house. The plain to the northeast of the lagoon (east of Tuineje) could also be of interest.

The tamarisks and palm trees on the northern edge of Gran Tarajal itself, and the Gran Tarajal Barranco in the area opposite the police station may also be worth a look, although development has reduced the value of this area in recent years.

Birds   Although the site is new, the list of species recorded here is already impressive, and includes the first Ruddy Shelduck for the Canary Islands (which subsequently bred), together with Spotted Crake, Collared Pratincole and Gull-billed Tern. Black-winged Stilts are frequent visitors, and may also have bred recently. The lagoon is particularly attractive to waders. Little Ringed and Kentish Plovers breed. Black-bellied Sandgrouse sometimes visit the area in mid-morning to drink. Swifts are usually present over the lagoon, and with luck, this is one place where Pallid and Plain Swifts can be seen together in summer.

The plain to the northeast supports both Cream-coloured Courser and Houbara Bustard, although it is not one of the better sites.

The tamarisk thicket south of Rosa de Catalina García holds good numbers of Sardinian Warbler, which also occurs around the cultivations, as well as in tamarisks in Gran Tarajal. Migrant passerines can occur in any of these areas at the right time. In winter the barranco in Gran Tarajal attracts birds such as Grey Wagtail (nominate race), Black Redstart, and a few waders.

Jandía (6)

Jandía is the southwest peninsula of Fuerteventura, and is virtually an island in its own right. It boasts the highest mountain in the Eastern Canary Islands (Pico de La Zarza, 807m) as well as the best beach in the whole of the Canary Islands. This last feature has given rise to rapid tourist development in recent years, and it is now one of the main tourist centres on the island. Fortunately for birdwatchers who stay in this area as part of a family holiday there are several good birdwatching areas nearby. The gardens surrounding the hotels and villas are good for migrants in spring and autumn. Planting and irrigation has now provided the closest approximation to woodland on the island since the laurel forests were destroyed in the fifteenth century. There are mountains and sandy plains in addition to the coastal habitats.

The Jandía peninsula also has a number of endemic plants, including the spectacular Jandìa Spurge, which is a large cactus-like plant now confined to a few small areas west of Morro Jable. The cliffs on the north side of the mountains also hold a range of rare plants. The rare endemic mat-forming Medusa's-head Bindweed grows on the sandy plain between the wind-farm and La Pared.

Location   The main road runs close to the south coast, but is metalled only as far as Morro Jable. The main birdwatching areas are around the base of the peninsula in the La Pared and Costa Calma area (see map page 74), and half way along the south coast around Punta de Matorral (see map page 75) with the resort gardens and mountains inland.

Strategy   The range of habitats on the peninsula is such that it holds virtually all the species of birds which occur in Fuerteventura, including all the island's specialities. It is also one of the best areas for migrant passerines and waders. It would, at least during the passage season, be quite possible to spend the whole of a week birdwatching on the peninsula. However, a holiday combining Jandìa with visits to other areas is likely to be more rewarding, at least for the first time visitor.

Jandía always seems to be a few degrees hotter than the other parts of the island, and the south coast is sheltered from the worst of the

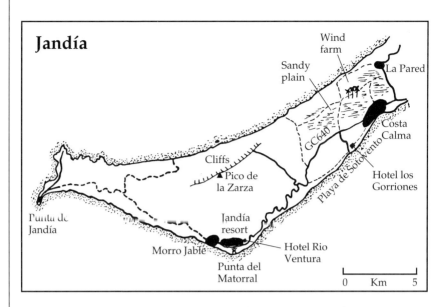

predominant northeasterly winds. For this reason, it is often important to make the most of the cooler mornings and evenings for active birdwatching. Early morning, when the sun-worshippers have yet to emerge, is the best time to check the beach areas. However, this is also the best time to be out on the sandy plains north of Costa Calma. The hotter part of the day can be used to check the wooded area, where there is good shade. Alternatively, take to the pool or the sea.

The best area of beach for waders is reached from the Los Gorriones Hotel, which is signposted from the main coast road 3km southwest of Costa Calma. The shore both sides of the hotel can be good, and there are shallow tidal lagoons which are quite attractive to waders and other birds. At low tide, the birds spread over a large area, and it is then difficult to cover, especially when the whole area is dotted with naturists! However, as the tide rises, the birds congregate nearer to the shore. At high tide, some of the waders move on to the sandy plains inland, or roost in quiet areas along the top of the low cliffs. In the early morning, before the heat haze starts, it is also possible to scan the tidal areas from the top of the low cliffs. The whole coast west to Punta de Matorral is of interest for waders.

Inland from the resort at this southern tip of the peninsula is the Pico de la Zarza. It can be climbed by taking the ridge path behind the Hotel Río Ventura. The climb is a hot and rather strenuous one, so go prepared with appropriate shoes, water, etc., if you attempt it. The views from the top are breathtaking, but beware if you do not have a head for heights as the sheer cliffs on the north side are very high.

The best area of sandy plain is to the east of the wind-farm, approximately half way between the two coasts. The area is fenced, but there appears to be no problem with access on foot. However, four-wheeled drive vehicles should not be taken into the area as this destroys the fragile habitat and may result in loss of nests or chicks. You can walk in from Costa Calma. Head inland from the wood, keep to the left of the tourist complex and park to the right of the farm

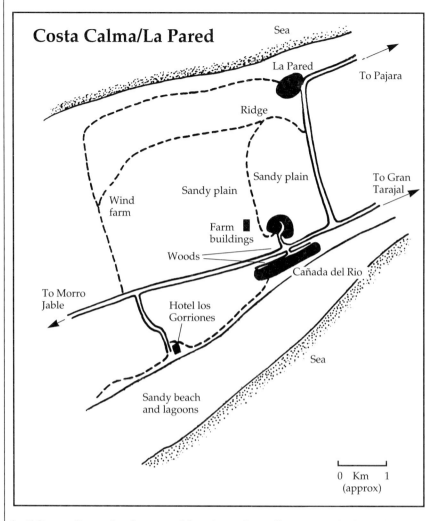

Costa Calma/La Pared

Sea

La Pared

To Pajara

Ridge

Sandy plain

To Gran
Tarajal

Sandy plain

Wind
farm

Farm
buildings

Woods

Cañada del Rio

To Morro
Jable

Hotel los
Gorriones

Sea

Sandy beach
and lagoons

0   Km   1
(approx)

buildings. Cross the fence and head up the valley towards the ridge, keeping the wind-farm to the left. Alternatively, drive across the peninsula towards La Pared and then turn left on a track along the ridge which divides the southern side of the peninsula from the northern side. Continue to the fence and then proceed on foot, heading down the slope towards the wind-farm.

Passerine migrants are attracted to the gardens in the vicinity of the tourist resorts, e.g. at Stella Canaris and Los Gorriones. However, the best area is perhaps the 'wood' in the Cañada del Río area of Costa Calma (either side of the main coast road), particularly at the western end, where there is more undergrowth. If there are any migrants around, then this wood deserves a thorough search. The saline scrub (Salicornia) between the coast road and the lighthouse at Punta del Matorral can also be quite good for migrants, as can the gardens in the nearby tourist complex.

The Punta de Jandía itself is only reached after an arduous journey along about 20km of rough track. The point may well be good for seawatching, but this appears, as yet, to be an untested theory.

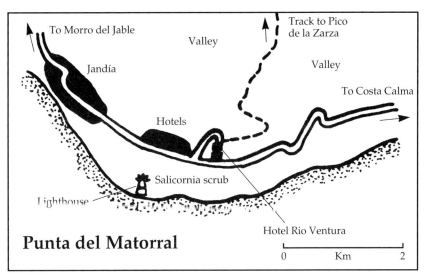

Punta del Matorral

0   Km   2

Birds   The sandy plain at the base of the Jandía peninsula, stretching from Costa Calma on the south coast, almost to La Pared on the north coast, is the best area in the Canary Islands for Black-bellied Sandgrouse. It is also the best site in the south of Fuerteventura for Houbara Bustard, and Cream-coloured Courser and Stone-curlew are common.

The sandy beaches hold good numbers of Kentish Plovers, and Little Egrets are usually present. Commoner waders such as Grey Plover,

Canary Island
Chat

Sanderling and Dunlin are frequent, and other species occur during passage including Black-winged Stilt, Curlew Sandpiper, and Bar-tailed and Black-tailed Godwits. The lagoons around Los Gorriones have regularly held the occasional Slender-billed Gull in recent years. This species has apparently bred in the past.

Jandía is probably the best area on the island for Barbary Falcon. It may nest in the Jandìa mountains, and the species is seen fairly regularly along the south coast. Until fairly recently, Ospreys were also seen along the coast, but these now appear to have died out on the island. Egyptian Vultures are still fairly common.

Passerine migrants have included Roller (near the lighthouse), while one visit to the wood at Costa Calma produced 3 Night Herons, 3 Wrynecks, Nightingale, Wood Warbler, and Pied Flycatcher, along with numerous Tree Pipits and Willow Warblers. In winter, the wood and other planted areas hold a few wintering passerines such as Song Thrush, Redwing, Robin, Blackcap and Chiffchaff.

The introduced Monk Parakeet now appears to be well established in the Stella Canaris area, and Collared Doves are also present.

Finally, the peninsula is a good place to see Canary Island Chat. The central mountains are a stronghold for the species, and it occurs almost to the very top of Pico de la Zarza. The track to the peak is a good place to look for it. However, it also occurs in many of the rocky barrancos lower down along the south side of the range (close to the main road), and can even be seen in the rockier areas behind the beach.

Caleta de Fustes (Castillo) and Barranco de la Torre (7)

The resort at Caleta de Fustes, now often referred to as El Castillo, is one of the three main holiday centres on the island, the others being Corralejo and Jandía. For the birdwatcher, it has two advantages over the other resorts; it is centrally placed for visiting all the key sites on the island, and there is reasonable birdwatching within four or five kilometres, so that a car is not absolutely essential (though still highly recommended).

Location　　Caleta de Fustes is on the east coast of Fuerteventura, about 7km south of the airport and 12km from Puerto del Rosario. The main areas for birdwatching in the vicinity are the gardens around the tourist complex itself, the plains to the southwest, and the Barranco de la Torre area, which is 4km south of the resort.

Strategy　　While the birdwatching around Fustes is not outstanding, there are a number of areas which are worth visiting for those birdwatchers who choose this as their base. With some luck, most of the island's specialities could be seen in the area in a week, though this is by no means guaranteed.

The gardens around the resort are attractive to migrants, and are always worth keeping an eye on in the periods March-May and August-October. The best area is the southwestern edge of the resort, around the Las Villas del Castillo complex which has good cover.

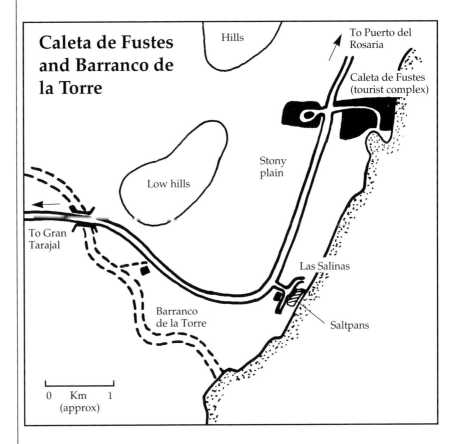

Caleta de Fustes and Barranco de la Torre

Hills

To Puerto del Rosaria

Caleta de Fustes (tourist complex)

Low hills

Stony plain

To Gran Tarajal

Las Salinas

Barranco de la Torre

Saltpans

0   Km   1
(approx)

Seawatching from the resort can be productive, especially when the prevailing northeasterly winds are particularly strong. Evenings are best since the site faces east.

To the southwest of the resort there is a stony coastal plain inland from the coast road. This is by no means one of the better plains on the island, but an extended early morning visit can be productive. In particular, check the areas further from the road, up to the edge of the low hills to the west.

On the coast itself, about 3km south of Fustes are the old salinas (salt-pans). These are reached by turning off the main road at the houses. While not extensive, the salt-pans do attract small numbers of waders, and the rocky shore to the north can also be quite productive. An early morning visit is recommended, since the waders are easily disturbed due to the small size of the pools.

Further south again is the Barranco de la Torre, which is perhaps the best area for general birdwatching in the vicinity. From Fustes, pass the turning to the salinas, follow the road round the bend, and look out for a track on the left which has a 'no entry' sign. Park here and follow the track on foot to the barranco. The barranco has a dense growth of tamarisk, and there is a stream along the bottom, which forms a series of pools of varying size, depending upon recent rainfall. Having reached the barranco, explore in both directions, checking the pools and bushes.

Birds   Lesser Short-toed Lark and Berthelot's Pipit are common on the stony plain, while Houbara Bustard, Cream-coloured Courser, and Black-bellied Sandgrouse could all be seen. Stone-curlew and Egyptian Vulture are fairly common in the area. Canary Island Chat breeds in the Barranco de la Torre and should be found at any time of the year. Sardinian Warblers are common in the tamarisks here, and Spectacled Warblers are particularly common around the barranco. Trumpeter Finches occur around the edges of the resort, and are also found in the barranco and around the salinas.

During migration, all the usual passerines are recorded, including species such as Nightingale, Blackcap, Willow and Wood Warblers, and Pied Flycatcher.

Little Ringed Plovers breed along the barranco, and migrant waders occur in small numbers. Grey Heron and Pochard have been seen on the pools. Small numbers of waders also occur at the salinas and Greater Flamingo has occurred. Species such as Little Egret, Whimbrel, and Grey Plover occur on the adjacent shore.

Seabirds offshore include the inevitable Cory's Shearwaters in summer, the occasional Gannet in winter, and Sandwich Terns all years. However, rarities have also been seen, including White-faced Storm-petrel, Little Shearwater and Red-billed Tropicbird.

Corralejo (8)

Corralejo is the main tourist resort at the northern end of Fuerteventura, and has the best tourist facilities on the island. It is handily located for some of the better birdwatching areas in the north, and also has some fairly good birdwatching close to hand. It is also the best base for a trip to Lanzarote, which is just six miles away by ferry.

Location   Corralejo is at the northeastern tip of the island, about 30km north of Puerto del Rosario.

Strategy   The area around Corralejo itself is not particularly good for birds, and a car is essential for anyone staying here and hoping to see most of the island's specialities. However, there are a few areas of interest. The shore is quite good for waders all year round, while the sandy plains just to the south hold many of the common species. During migration, the shrubs around the villa complexes can be good for migrants, although with the ever increasing number of shrubs to chose from, locating them is becoming more difficult. The isolated complex known as Parque Holandes, 11km down the coast road to the south, is perhaps the best area now, simply because the habitat is more limited.

Although the immediate area is not always productive, there are good sites nearby at Cotillo and La Oliva (see pages 65 and 67), and these can be reached by bus as well as by car.

It is possible to visit the island of Lobos, which is a volcanic cone just offshore. This island is a major breeding site for Cory's Shearwater and possibly other seabirds, although nothing can be seen during the day. A trip over the channel to Lanzarote may also be worthwhile. A hire car can be taken over for about £50, and it is then easy enough to visit sites

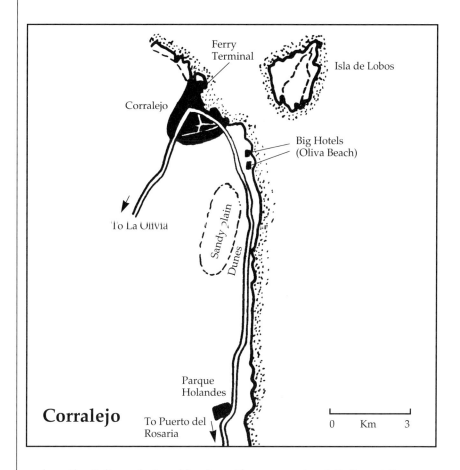

Corralejo

such as the Salinas de Janubio, Arrecife, or even to visit sites at the northern end of the island. However, two days should be allowed if you plan to cover the island more thoroughly. The ferry crossing is not particularly good for seabirds, but it can produce good birds from time to time, and it is certainly worth keeping your eyes open, particularly during an evening crossing.

Birds   The shores to the south and west of the town hold waders such as Kentish Plover and Whimbrel, and Little Egrets are fairly common. Cory's Shearwaters breed on Lobos, and these occur in large numbers offshore in the evenings. They can be seen well from the ferry crossing to Lanzarote, and other birds which have recently been seen from the ferry include Red-billed Tropicbird, Pomarine Skua and Black Tern.

The dune area to the south of Corralejo and inland from the Oliva Beech Hotel complex, is rather devoid of birds, although Stone curlew occurs. Part of the area is designated as a National Park, and there are rather amusing roadside signs in the form of Houbara Bustards. While bustards used to occur in good numbers, most birdwatchers fail to see them here now, and it cannot be recommended as a good site. The areas of consolidated sand towards the rear of the dune complex (i.e. close to the lava fields) are the best areas to search, particularly at the southern end. Berthelot's Pipit and Spectacled Warbler are also common in this area.

Other resident species which should be found within walking distance of the town include Pallid Swift, Lesser Short-toed Lark, Hoopoe, Great Grey Shrike, and Trumpeter Finch. Spanish Sparrows are common.

Migrants which have been recorded include scarcer species such as Bee-eater, Wryneck, Tawny Pipit, Melodious and Bonelli's Warblers, and Ortolan Bunting. Rarities such as Great Spotted Cuckoo, Olivaceous and Yellow-browed Warblers, and Brambling have also been found here.

Barranco de Río Cabras
(Willis's Barranco or Airport Barranco) (9)

This is one of the largest barrancos along the east coast of the island. The barranco is narrow and deep, with cliffs on either side. It is normally completely dry. Dry, rocky plains and low hills occur to either side. There is also a small, usually dry reservoir. The barranco is a fairly reliable site for Canary Island Chat.

Location   The barranco is just a few hundred metres north of the airport, and is

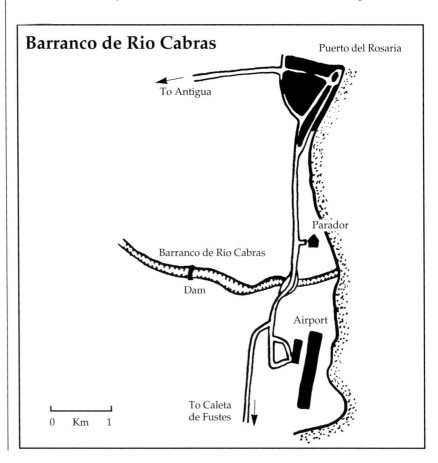

crossed near its seaward end by the main road linking the airport with Puerto del Rosario.

The road crosses the barranco on a high embankment, and it is possible to pull off onto the old road, where it is safe to park. Alternatively, continue on past the barranco and park off the road near the Autos Dominguez garage.

Strategy

This site is really only worth visiting if you are on a flying visit to Fuerteventura, and wish to see Canary Island Chat as close to the airport as possible. Indeed, it has been known for twitchers to fly to Fuerteventura, walk to the barranco, and fly out again a few hours later with the island's only endemic bird species safely ticked off.

Walk inland along the floor of the barranco, checking tops of shrubs and other vegetation for the chat. The reservoir is about a mile and a half up the barranco, but the chat should be seen well before this, and it is not normally worth going as far as the reservoir since it is almost always dry. It is also possible to climb out onto the plains on either side, and these could be worth checking in early morning.

Birds

In addition to the chat, there are Pallid and perhaps Plain Swifts breeding in the cliffs. Barbary Partridge and Trumpeter Finch also occur in the barranco. Black-bellied Sandgrouse are often seen here, and both Houbara Bustard and Cream-coloured Courser have been recorded on the plains in the vicinity.

Trumpeter Finch

LANZAROTE

Birds to be seen in Lanzarote are much the same as in Fuerteventura, although Black-bellied Sandgrouse and Canary Islands Chat are absent. Barbary Falcon occurs, and Eleonora's Falcon nests on offshore islands. Houbara Bustards are fairly common, although suitable habitat is more restricted than on Fuerteventura.

Playa Blanca (1)

Playa Blanca is a major tourist centre. New tourist complexes are being built all the time, but it does still retain a rather quieter feel than resorts further north, and it can be recommended as a good base for a birdwatching holiday.

Location   Playa Blanca is at the extreme southern end of Lanzarote, looking across the Straights of Bocaina to Fuerteventura. The town is therefore well-placed for a day trip to Fuerteventura. It is also convenient for the salinas (salt-pans) at Janubio.

Strategy   In the appropriate seasons the gardens associated with the hotels and villa complexes are highly attractive to migrants, and the more densely-planted areas should be thoroughly checked if there is any indication of

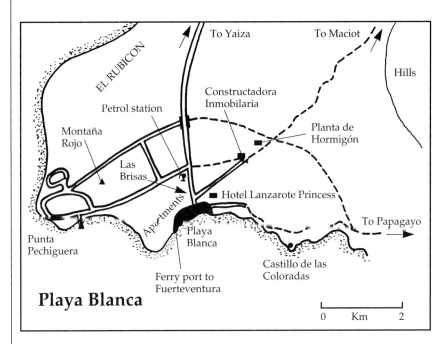

Playa Blanca

migration. The vegetation around the Las Brisas apartments and the
Hotel Lanzarote Princess are among the best areas for migrants.
During the early part of the morning concentrate on the plains to the
north of the town. It is best to search those areas furthest from the
tourist developments, either to the west of the main road going north,
in the area known as El Rubicín, or to the east of this road and west
of the minor (unsealed) road towards Maciot. Seawatching from
Punta Pechiguera can be worthwhile in the early mornings and
evenings.

Birds   Houbara Bustards and Cream-coloured Coursers occur on the plains,
although the bustard doesn't seem to be as numerous in the area as it
once was, presumably due to the continued rapid development which
has taken place over the last few years. Other resident species which
can be seen on the plains include Stone-curlew, Lesser Short-toed Lark,
Berthelot's Pipit, Spectacled Warbler, Great Grey Shrike and Trumpeter
Finch. Cory's Shearwaters are numerous off Punta Pechiguera from
March to November. Scarcer migrants and rarities which have been
seen around the town include Booted Eagle, Marsh, Montagu's and
Pallid Harriers, Golden Oriole, and Subalpine and Bonelli's Warblers,
while commoner migrants such as Blackcap, Willow Warbler, and Pied
Flycatcher can be quite numerous.

Salinas de Janubio and Laguna de Janubio (2)

The large area of salt-pans known as the Salinas de Janubio, together
with its tidal lagoon, is one of the better known birdwatching sites
on the island, primarily for waders. Unfortunately, the salt-pans
are no longer used, and the attractiveness of the area for birds appears
to have declined as a result. Nevertheless, the area is still worth a visit.

Salina de Janubio/Laguna de Janubio

To Yaiza

Bridge

Las Breñas

Lava fields

Saltpans

Layby

To El Golfo

Laguna de Janubio

R Restaurant

P Car parking areas

0   Km   1
(approx)

To Playa Blanca

Location
    The salinas are 13km north of Playa Blanca on the west coast of the island, at the southern flank of the lava fields. The entrance to the salt-pans is 400m from the main Playa Blanca-Yaiza road. It is also possible to park at the seaward end of the lagoon, which is another 1.7km further on, or at a layby above the salt-pans midway between this car park and the nearby restaurant. The restaurant itself has a terrace overlooking the salt-pans.

Strategy
    The salt-pans are worth visiting at any time of year. Access to the area is unrestricted. The waders frequent the shallow pools, as well as the shore of the adjacent tidal lagoon

Birds
    Kentish and Little Ringed Plovers breed at the salinas. A number of other waders occur in winter and on migration, including species such as Greenshank, Common Sandpiper, Little Stint, Curlew Sandpiper, and Bar-tailed and Black-tailed Godwit. Little Egret, Barbary Falcon and Osprey are also regular. Rarer species have occurred, including Greater Flamingo, Purple Heron, Black-winged Stilt, and Avocet. Black-necked Grebe has been seen on the tidal lagoon.

Arrecife Harbour and Charco (3)

Arrecife is the largest town in the Eastern Canary Islands. The shore near the centre of town is good for waders.

Strategy   The main birdwatching areas are all close to the city centre. View the shore from Avenida Generalísimo Franco. It is then worth crossing the bridge and walking out to the Castillo de San Gabriel to check the shores of the island on which it stands. Further along the promenade a bridge crosses the mouth of a tidal lagoon known as the Charco de San Gines. This is a good place for waders, gulls and egrets. It is possible to walk round the southern part of the lagoon. Further along is the harbour, which can also be worth checking.

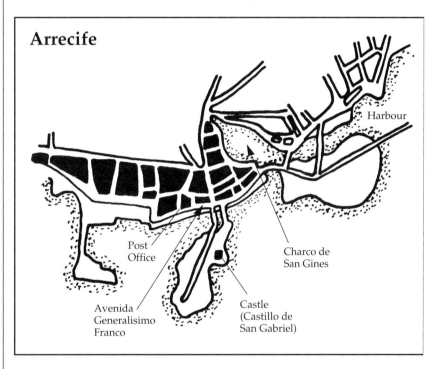

Arrecife

Harbour

Post Office

Charco de San Gines

Avenida Generalisimo Franco

Castle (Castillo de San Gabriel)

Birds   The rocky shores and offshore reefs attract a wide variety of waders such as Ringed and Grey Plovers, Whimbrel, Turnstone and Common Sandpiper. The area also attracts herons, gulls, and terns. Little Egrets and Grey Herons are usually present, and Sandwich Terns are common. During passage a variety of other waders occurs, and a number of rarities have occurred in the area. These have included Spotted Sandpiper and Whiskered Tern. Cattle Egrets have nested recently in the palm trees along Avenida Generalísimo Franco.

Riscos de Famara (4)

The cliffs of the Riscos de Famara are one of the natural spectacles of the island. They stretch for about 10km and are up to 600m (2,000 feet) high. The cliffs are the only reliable site in the Eastern Canary Islands for Barbary Falcon and also attract other birds of prey sometimes including Eleonora's Falcon. Several of the plants which can be found on the cliff-tops are endemic to the Famara region.

Location   The cliffs are at the northern end of Lanzarote, overlooking the island of Graciosa. There are viewpoints at Mirador del Río near the northern tip of the island, and at Guinate which is about half way between Mirador del Río and Haría. The Mirador de Guinate is beyond the village near the Tropical Park.

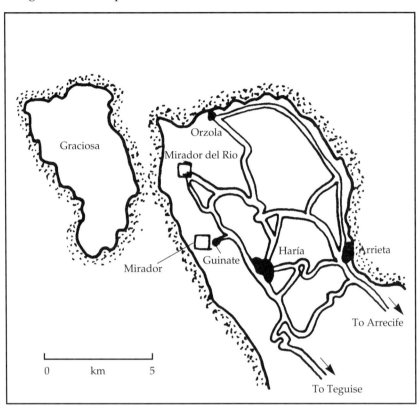

Strategy   Park at either of the miradors and scan the cliffs.

Birds   Barbary Falcons can be seen from either site. In summer (May-September) Eleonora's Falcon, which nests on the islets to the north of Lanzarote may be an occasional visitor to the cliffs. Other raptors which may be seen here include Osprey and Egyptian Vulture. Apart from the birds of prey there is little more to be seen than Yellow-legged Gulls and Rock Doves. However, Greenfinch and Corn Bunting (both very scarce in the eastern islands) occur in the area.

Teguise Plain/El Jable (5)

This is an extensive plain, part of which is under cultivation, and part of which is sandy waste. The plain is one of the best sites on Lanzarote for Houbara Bustard and most of the other desert species occur here as well.

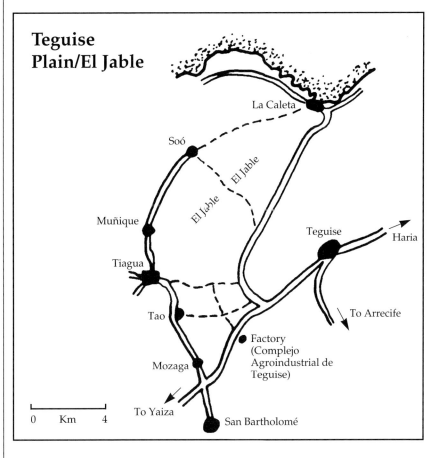

Teguise
Plain/El Jable

La Caleta

Soó

El Jable

El Jable

Muñique

Teguise

Haria

Tiagua

Tao

To Arrecife

Factory
(Complejo
Agroindustrial de
Teguise)

Mozaga

0   Km   4

To Yaiza

San Bartholomé

Location

The plain is near the centre of the island, to the south and west of the town of Teguise. From Teguise take the road towards Yaiza, and after 3km turn right towards La Caleta. If approaching from Yaiza, pass a large factory on the right (Complejo Agroindustrial de Teguise) and take the next road to the left. The plain is to the west of this road. The southern end of the plain is partially cultivated, whereas the area towards the coast (El Jable) is predominantly uncultivated and appears rather less productive for birds.

Strategy

The area is best visited early in the morning when the plains birds are more active, and before the heat haze makes birdwatching difficult. The easiest way to cover it is to drive along the numerous tracks which cross the plain, stopping occasionally at suitable vantage points to scan the cultivations and open plains. However, the site can also be covered on foot.

The best area appears to be about 5km2 in extent, and is reached by a series of tracks on the left side of the first 2.5km of the La Caleta road. The track to Tao is reached after 0.5km, the track to Tiagua after another 0.8km, and the track to Sío after another 1.2km.

Birds

Houbara Bustard, Cream-coloured Courser and Stone-curlew all occur on the plain in good numbers. It is also a good area for Lesser

Short-toed Lark, Hoopoe, Berthelot's Pipit, and Great Grey Shrike. Other species include Spectacled Warbler, Raven, and Trumpeter Finch.

Houbara Bustard

Tahiche Golf Course (6)

Tahiche Golf Course is the only large area of irrigated grassland in the Eastern Canary Islands. As such it is an attractive area for both resident and migrant birds.

Location The golf course is inland from Arrecife and the resort of Costa Teguise. From Costa Teguise, there is a signpost to the 'Campo de Golf' from the Avenida de las Palmeras (Lanzarote Bay Hotel) at the southern end of the resort, and the golf course is 3km distant.

Strategy The site is best during the migration seasons. Entry to the golf course is restricted, and there is a fence to prevent unauthorised access. Ask for permission to enter the golf course at the club house. However, there are roads from which it is possible to view much of the course area, even if entry is refused. The car park by the entrance also provides views over much of the course.

Birds Resident birds which can be seen in the area include Stone-curlew, Hoopoe (common), and Great Grey Shrike. Moorhen, which is scarce on this island, also occurs here. Barbary Partridge occurs both on and around the golf course, and this is one of the easiest places on the island to see this species.

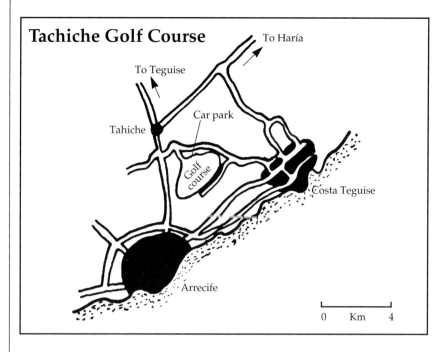

Tachiche Golf Course

To Haría
To Teguise
Car park
Tahiche
Golf course
Costa Teguise
Arrecife
0   Km   4

During the migration season, the large expanse of grass, together with associated trees and shrubs, are attractive to a wide range of migrants. The area is perhaps particularly attractive to Bee-eaters, and other species which have been recorded include Purple Heron, Cattle Egret, Woodchat Shrike, and Bonelli's Warbler.

Los Cocoteros Salt-Pans (7)

With the partial demise of the salt-pans at Janubio, the small area of salt-pans at Los Cocoteros is now the best place in Lanzarote for migrant waders.

Location   The salt-pans are on the east coast about 3km from Guatiza village. Enter Guatiza on the GC710, heading north from Arrecife, and turn right near the entrance to the village along Calle Tajasnayo. Follow the rather twisty road towards Los Cocoteros village, and after 2.4km turn left towards houses. After another 0.5km there is a small parking area on the right, beside a row of houses. The salt-pans are over a series of stone walls to the rear of the houses. Alternatively, continue straight on at the last junction. The road becomes a dirt track after a few hundred metres and bears to the left, ending at the shore.

Strategy   From the entrance by the houses, approach the salt-pans carefully, using the stone walls as cover. With care, the waders can be viewed at close range. From the seaward end, there are slightly more distant views over most of the pools. It is also possible to walk north along the shore from here, with the salt-pans to the left. View from both sides of the salt complex to ensure complete coverage.

Birds   A good range of waders can be seen especially during migration.

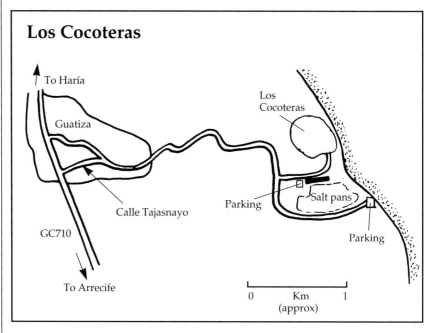

Los Cocoteras

To Haría

Los Cocoteras

Guatiza

Calle Tajasnayo

Parking

Salt pans

Parking

GC710

To Arrecife

0   Km   1
(approx)

Species regularly recorded include Little Stint, Common Sandpiper, Ringed, Little Ringed and Kentish Plovers, Black-tailed and Bar-tailed Godwits, Turnstone, etc. Other species which have been observed include Teal, Mallard, Black-winged Stilt and Curlew Sandpiper.

Mirador de Haria (8)

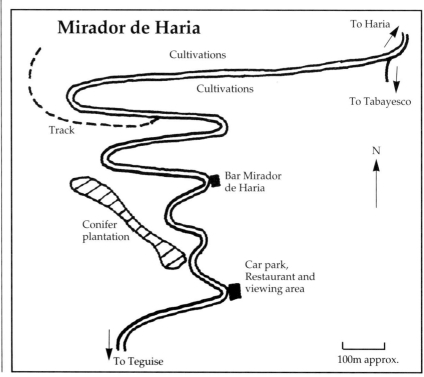

Mirador de Haria

To Haria

Cultivations

Cultivations

To Tabayesco

Track

N

Bar Mirador de Haria

Conifer plantation

Car park, Restaurant and viewing area

To Teguise

100m approx.

The area around the Mirador de Haria has extensive cultivations, scattered conifer plantations, and areas of scrub vegetation containing a number of very local endemic plants. It has recently become a reliable site for Canary.

Location

The town of Haria is in the northern part of Lanzarote and nestles among the mountains of the Famara Massif. The main road south from Haria (towards Teguize) winds up in to the mountains through a cultivated area with large trees and shrubs. Two view points can be found beside the road (see map).

Strategy

Drive south from Haria and park by the road overlooking the cultivations on the right, just before the first hair-pin bend, and scan for Canaries. Alternatively, drive down the track on the right-hand side of the road, just after the bend, and walk down into the cultivated areas. Canaries can also be seen from the lower of the two view points (i.e. the closest to Haria) and near the conifers above this point.

Birds

During the last few years this has become a reliable site for Canary, though whether the birds arrived naturally or were introduced is not known. Other birds to be seen in the area include Kestrel, Raven and Berthelot's Pipit.

Little Shearwater

SELECTED BIRD SPECIES

The following list provides additional information about a selection of local or uncommon species which can be difficult to find. Each of the endemic species are also mentioned.

Bulwer's Petrel A breeding summer visitor, usually seen between early May and September. It is most often seen from the ferry between Tenerife and Gomera, but can also be seen by seawatching from certain headlands on Tenerife and possibly from other islands.

Little Shearwater This species is recorded in most months of the year, but seems to be more common in the autumn. The Tenerife to Gomera ferry provides a very good chance of seeing this species. It can also be seen from virtually any headland in the Canary Islands, although it is only seen very occasionally in the eastern islands.

Madeiran Storm-petrel A rare breeding winter visitor, usually seen in August, but also recorded throughout the autumn and in May.

Marbled Duck A rare bird which has recently been seen on a regular basis at Los Molinos reservoir on Fuerteventura, but rarely recorded elsewhere.

Egyptian Vulture Now virtually confined to Fuerteventura, where it is still reasonably common, and Lanzarote where it is rather scarce.

Eleonora's Falcon This species arrives in May and departs in September. It nests on the islets north of Lanzarote, and the best place to see it is from the Famara cliffs where the species probably occurs fairly regularly in summer.

Barbary Falcon The Famara cliffs on Lanzarote is one of the best place to see this rare species. The other reliable site is the cliffs at Teno in northwest Tenerife. It can occur anywhere in the islands, but beware of confusion with wintering Peregrines.

Barbary Partridge This species is relatively common on Fuerteventura, where any rocky area can be productive. The best time to see it is just before the shooting season starts in early September, when there are flocks everywhere. In spring, the more inaccessible areas should be searched. The mountains around Las Peñitas are particularly good. In Lanzarote, the area around Tahiche Golf Course seems to be the best. On Tenerife, the arid southern part of the island is best.

Houbara Bustard Although the Houbara Bustard has a large world range, it is very difficult to find in most countries. It is much commoner on the Eastern Canary Islands than anywhere else. The highest density of this species is probably on the plain southwest of Teguise on Lanzarote. The other area on Lanzarote where it is regularly seen is just north of Playa Blanca, at the extreme southern end of the island. However, the real home of the Houbara Bustard in the Canary Islands is Fuerteventura, where there are several hundred individuals. It can be found on just about any of the flatter areas, especially those where there is relatively well-developed scrub vegetation, or where there are cultivations nearby. Although it is fairly common, it is quite hard to see because of its elusive behaviour. An early start is a must when searching for this species – the first two or three hours of daylight are best, and it is much harder to find between mid-morning and mid-afternoon. Also, it is much easier to get decent views from a vehicle. For this reason, the plains which have driveable tracks are the best places to look (e.g. Teguise on Lanzarote, and Cotillo on Fuerteventura).

Cream-coloured Courser This bird is found on Fuerteventura and Lanzarote. It is fairly common on both islands, and is found in much the same areas as Houbara Bustard, so that it will almost always be seen during searches for that species.

Black-bellied Sandgrouse Fuerteventura is the only island where this species normally occurs, although it does wander to Lanzarote from time to time. The main centre of the population is the sandy plain north of Costa Calma, where the species can almost be guaranteed. Catalina García lagoon is also a good place to see the bird. Smaller numbers occur throughout the island. The loud bubbling flight call is often the first indication of its presence.

Bolle's Pigeon This is the commoner of the two endemic pigeons, and is by far the easiest to see on Tenerife. Flight views are relatively easy to obtain (given good weather) at a variety of laurel forest sites, although there are fewer places where they are regularly seen perched.

Laurel Pigeon This species is scarce on Tenerife, where flight views are usually all that can be obtained. However, there are better sites on Gomera and La Palma where they are commoner and easier to see.

Plain Swift This endemic species is by far the commonest species of swift on Tenerife and the western islands. However, it is scarce in the eastern islands where it tends to be associated with deep barrancos. Care is needed in identifying the species since confusion with Pallid Swift is a possibility under certain light conditions.

Pallid Swift The common swift species on the eastern islands. More difficult to see in the other islands.

Great Spotted Woodpecker The endemic subspecies occurring in the native pine forests on Gran Canaria and Tenerife. It is fairly common and confiding on Gran Canaria, and also occurs at several of the Blue Chaffinch sites on Tenerife.

Lesser Short-toed Lark The endemic sub-species is abundant on Fuerteventura and Lanzarote, but has a more restricted range on Gran Canaria, and is confined to the south on Tenerife where it is a scarce inhabitant of arid plains. The nominate race is confined to the La Laguna area on northern Tenerife.

Berthelot's Pipit This Macaronesian endemic is so common (in all habitats except closed woodland, and on all islands) that it is invariably seen every day.

Red-throated Pipit A very scarce winter visitor to Tenerife (and other islands?). The best sites are Golf del Sur, Amarillo Golf and Los Palos in the south of Tenerife.

Sardinian Warbler A common bird in the lusher western islands, but confined to certain habitats in the eastern islands, where it is primarily found in Tamarisk thickets. In Fuerteventura, shrubs in Betancuria and the public park in Pájara are also good places to see it.

Tenerife Goldcrest Primarily a bird of mixed forest particularly in the middle latitudes along the northern side of Tenerife. Also found on the other western islands.

Canary Islands Chat Confined to Fuerteventura, where it is local, being fairly common in the mountainous centre of the island and in Jandía. The roadside between Pájara and Vega de Río Palmas is a good site, while there are several barrancos along the east coast which are also good.

Rock Sparrow In the Canary Islands, this is a localised bird of high altitude villages and ravines. It is absent from Fuerteventura and Lanzarote. It is relatively common in Gran Canaria, particularly around Atlantera. It is also found on Tenerife, where it is best looked for in Vilaflor and around Santiago del Teide.

Blue Chaffinch Found on Tenerife and Gran Canaria only, and restricted to native pine forests. The species is much commoner on Tenerife, where there are a number of good sites. There are a number of water pipes where good views can often be had. On Gran Canaria the species is very scarce, and there is really only one good site.

Canary A common bird on Tenerife and the other western islands, but rather local on Gran Canaria, where it is commonest in cultivated and wooded areas at middle altitudes. Also now found at one site on Lanzarote.

Trumpeter Finch Fairly common on Lanzarote and Fuerteventura, occurring throughout both islands. Attracted to water, and easiest to see at reservoirs, wells and other water sources, but sometimes common round the edges of villages where there are cultivations. On Tenerife and Gran Canaria it is a bird of the arid southern areas.

ASHMOLE, M. and P. ASHMOLE. 1989. Natural history excursions in Tenerife: a guide to the countryside, plants and animals. Kidston Mill Press, Scotland.

ASKEW, R. R., 1988. The Dragonflies of Europe. Harley Books, Colchester.

BANNERMAN, D. A. 1922. The Canary Islands: their history, natural history and scenery. Gurney and Jackson, London and Edinburgh.

BANNERMAN, D. A. 1963. Birds of the Atlantic Islands volume 1: a history of the birds of the Canary Islands and of the Salvages. Oliver and Boyd, Edinburgh and London.

BARBADILLO, L.J. 1987. La Guía de Incafo de los Anfibios y Reptiles de la Peninsula Iberica, Islas Baleares y Canarias. Pub. Guias Verdes de Incafo, Madrid.

BRAMWELL, D. and Z. I. BRAMWELL. 1974. Wild Flowers of the Canary Islands. Stanley Thornes (Publishers) Ltd., London and Burford.

BRAMWELL, D. and Z. I. BRAMWELL. 1987. Historia Natural de las Islas Canarias – Guia Basica. Editorial Rueda, Madrid.

COLLAR, N. J. and S. N. STUART. 1985. Threatened Birds of Africa and Related Islands. International Council for Bird Preservation, Cambridge.

CONCENPIÓN GARCÍA, D. 1992. Avifauna del Parque Nacional de Timanfaya. Instituto Nacional para la Conservation de la Naturaleza.

CRAMP, S. et al. 1977-1994. Handbook of the Birds of Europe the Middle East and North Africa – the birds of the Western Palearctic, volumes 1-9. Oxford University Press, Oxford.

FERNANDEZ, J. M. 1978. Los lepidopteros diurnos de las Islas Canarias. Enciclopedia Canaria. Aula de Cultura de Tenerife.

FERNANDEZ-RUBIO, F. 1991. Guía de Mariposas Diurnas de la Peninsula Iberica, Baleares, Canarias, Azores y Madeira (2 vols) Pub Pirámide, Madrid

HIGGINS, L. G. and B. HARGREAVES. 1983. The Butterflies of Britain and Europe. Collins, London.

HEINZEL, H., R. FITTER and J. PARSLOW. 1995. Collins Pocket Guide to Birds of Britain and Europe, with North Africa and the Middle East. 5th Edition. HarperCollins, London.

Selected bibliography

KUNKEL, G. (Ed). 1976. Biogeography and ecology in the Canary Islands. W.Junk.B.V., The Hague.

LORENZO GUTIÉRREZ, J. A., and J. González Domínguez. 1993. Las Aves de Médano. Asosiacion Tinerfeña de Amigos de la Naturaleza.

MARTIN, A. 1987. Atlas de las Aves Nidificantes en la Isla de Tenerife. Instituto de Estudios Canarios, Tenerife.

MORENO, J. M. 1988. Guía de las Aves de las Islas Canarias. Editorial Interinsular Canaria.

PEREZ PADRON, F. 1986. The birds of the Canary Islands. Enciclopedia Canaria. Aula de Cultura de Tenerife. 3rd edition.

ROCHFORD, N. 1984. Landscapes of Tenerife – a countryside guide. Sunflower Books, London.

ROCHFORD, N. 1986. Landscapes of Gran Canaria – a countryside guide. Sunflower Books, London.

ROCHFORD, N. 1989. Landscapes of Lanzarote and Fuerteventura – a countryside guide. Sunflower Books, London.

WATSON, L. 1985. Whales of the World – a handbook and field guide to all the living species of whales, dolphins and porpoises. Hutchinson, London.

FULL SPECIES LIST

The following list includes all birds mentioned in Moreno (1988), together with all additional species known to have occurred by the authors. The total is 350, of which one is extinct. A large proportion of these are either passage migrants or vagrants. It should be noted that some of the species listed have not been submitted to the local rarities committee, and some may refer to escapes.

The range of species to be seen is different for each of the islands, and some species are common on some islands but uncommon on others. For resident species, this is due to the fact that not all islands have a complete range of habitats. For migrants it is caused by the fact that the easternmost islands are closer to the African coast, and therefore tend to receive larger numbers of migrants. This means that the visitor needs to know the status of each species on each island, and we have tried to indicate this using a series of symbols. We have also provided details on the status of each of the islands' endemic species and sub-species.

We would like to encourage all visitors to the islands to submit their records to Tony Clarke at the following address: C/República Dominicana, No 61, Barrio de Fátima, 38500 Güimar, Tenerife, Spain. Tel. 52-42-91

The following abbreviations and symbols are used:

Status

R	=	Resident
S	=	Summer visitor
B	=	Breeds
(b)	=	Has bred
W	=	Winter visitor
M	=	Passage migrant
(E)	=	Migrants mainly recorded from the Eastern Islands
(W)	=	Migrants mainly recorded from the Western Islands
A	=	Accidental
I	=	Introduced resident
1	=	Abundant
2	=	Common
3	=	Fairly common
4	=	Uncommon
5	=	Very localised or rare
?	=	Status uncertain
(s)	=	ship assisted

Islands

H	=	El Hierro
G	=	La Gomera
P	=	La Palma
T	=	Tenerife
GC	=	Gran Canaria

F	= Fuerteventura
L	= Lanzarote
I	= Islets north of Lanzarote

Endemics

ec	= Canary Islands endemics
em	= Macaronesian endemics
*	= Subspecies confined to Macaronesia
**	= Subspecies confined to the Canary Islands

A	☐	Black-throated Diver (Gavia arctica)
A	☐	Little Grebe (Tachybaptus ruficollis)
A	☐	Great Crested Grebe (Podiceps cristatus)
W5	☐	Black-necked Grebe (Podiceps nigricollis)
A	☐	Pied-billed Grebe (Podilymbus podiceps)
A	☐	Albatross sp. (Diomedea ? melanophris)
A	☐	Fulmar (Fulmarus glacialis)
emA	☐	Soft-plumaged Petrel sp. probably Fea's Petrel (Pterodroma feae)
SB4	☐	Bulwer's Petrel (Bulweria bulwerii) H,T,I
*SB1	☐	Cory's Shearwater (Calonectris diomedea borealis)
M(W)4	☐	Great Shearwater (Puffinus gravis)
A	☐	Sooty Shearwater (Puffinus griseus)
M(W)3(b)	☐	Manx Shearwater (Puffinus puffinus) breeds T,P,G?,H?
A	☐	Mediterranean Shearwater (Puffinus yelkouan)
*R/WB4	☐	Little Sheatwater (Puffinus assimilis baroli) H,T,GC,I
A	☐	Wilson's Storm-petrel (Oceanites oceanicus)
SB5	☐	White-faced Storm-petrel (Pelagodroma marina) I
SB4	☐	Storm-petrel (Hydrobates pelagicus) H,T,I
A	☐	Leach's Storm-petrel (Oceanodroma leucorhoa)
WB5	☐	Madeiran Storm-petrel (Oceanodroma castro) off T, I
A	☐	Red-billed Tropicbird (Phaethon aethereus)
W3	☐	Gannet (Sula bassana)
W5	☐	Cormorant (Phalacrocorax carbo)
A	☐	Shag (Phalacrocorax aristotelis)
A?	☐	Pink-backed Pelican (Pelecanus rufescens)
A	☐	Bittern (Botaurus stellaris)
A	☐	American Bittern (Botaurus lentiginosus)
M4	☐	Little Bittern (Ixobrychus minutus)
A	☐	Dwarf Bittern (Ixobrychus sturmii)
M4	☐	Night Heron (Nycticorax nycticorax)
A	☐	Squacco Heron (Ardeola ralloides)
W/M4(b)	☐	Cattle Egret (Bubulcus ibis) breeds L
A	☐	Western Reef Heron (Egretta gularis)
R3	☐	Little Egret (Egretta garzeta)
A	☐	Great White Egret (Egretta alba)
R3	☐	Grey Heron (Ardea cinerea)
M4	☐	Purple Heron (Ardea purpurea)
A	☐	Black Stork (Ciconia nigra)
M5	☐	White Stork (Ciconia ciconia)
A	☐	Glossy Ibis (Plegadis falcinellus)
A	☐	Sacred Ibis (Threskiornis aethiopicus)

W/M4 ☐	Spoonbill (Platalea leucorodia)
A ☐	Greater Flamingo (Phoenicopterus ruber)
A ☐	Bean Goose (Anser fabilis)
A ☐	Pink-footed Goose (Anser brachyrhynchus)
A ☐	White-fronted Goose (Anser albifrons)
A ☐	Greylag Goose (Anser anser)
A ☐	Brent Goose (Branta bernicla)
A ☐	White-faced Whistling Duck (Dendrocygna viduata)
A ☐	Shelduck (Tadorna tadorna)
A(b) ☐	Ruddy Shelduck (Tadorna ferruginea)
W4 ☐	Wigeon (Anas penelope)
A ☐	American Wigeon (Anas americana)
A ☐	Falcated Duck (Anas falcata)
W5 ☐	Gadwall (Anas strepera)
W3 ☐	Teal (Anas crecca crecca)
A ☐	Green-winged Teal (Anas crecca carolinensis)
W4 ☐	Mallard (Anas platyrhynchos)
A ☐	Black Duck (Anas rubripes)
W4 ☐	Pintail (Anas acuta)
M4 ☐	Garganey (Anas querquedula)
A ☐	Blue-winged Teal (Anas discors)
W4 ☐	Shoveler (Anas clypeata)
A(b) ☐	Marbled Duck (Marmaronetta angustirostris)
A ☐	Red-crested Pochard (Netta rufina)
W3 ☐	Pochard (Aythya ferina)
W4 ☐	Tufted Duck (Aythya fuligula)
A ☐	Ring-necked Duck (Aythya collaris)
A ☐	Ferruginous Duck (Aythya nyroca)
A ☐	Scaup (Aythya marila)
A ☐	Lesser Scaup (Aythya affinis)
A ☐	Common Scoter (Melanitta nigra)
A ☐	Red-breasted Merganser (Mergus serrator)
A ☐	Honey Buzzard (Pernis apivorus)
M4 ☐	Black Kite (Milvus migrans)
A(EB) ☐	Red Kite (Milvus milvus)
A ☐	White-tailed Eagle (Haliaeetus albicilla)
RB3 ☐	Egyptian Vulture (Neophron percnopterus) GC?,F,L
M/W4 ☐	Marsh Harrier (Circus aeruginosus)
M/W4 ☐	Hen Harrier (Circus cyaneus)
A ☐	Pallid Harrier (Circus macrourus)
M4 ☐	Montagu's Harrier (Circus pygargus)
*RB3 ☐	Sparrowhawk (Accipiter nisus granti) H,P,G,T
**RB3 ☐	Buzzard (Buteo buteo insularum) H,P,G,T,GC,F(4),L(5),I
A ☐	Golden Eagle (Aquila chrysaetos)
A ☐	Booted Eagle (Hieraaetus pennatus)
A ☐	Bonelli's Eagle (Hieraaetus fasciatus)
RB ☐	Osprey (Pandion haliaetus) H,G,T,GC,F(E?),L,I
A ☐	Lesser Kestrel (Falco naumanni)
*RB1 ☐	Kestrel (Falco tinnunculus canariensis) H,P,G,T,GC
**RB1 ☐	(Falco tinnunculus dacotiae) F,L,I

A ☐	Red-footed Falcon (Falco vespertinus)
A ☐	Merlin (Falco columbarius)
M4 ☐	Hobby (Falco subbuteo)
RB5 ☐	Eleonora's Falcon (Falco eleonorae) L?,I
A ☐	Lanner (Falco biarmicus)
W4 ☐	Peregrine Falcon (Falco peregrinus)
RB5 ☐	Barbary Falcon (Falco pelegrinoides) T,GC,F,L,I
IRB3 ☐	Red-legged Partridge (Alectoris rufa) GC
RB3 ☐	Barbary Partridge (Alectoris barbara) P,G,T,F,L
*RB4 ☐	Quail (Coturnix coturnix) H,P,G,T,GC,F,L
IRB5 ☐	Helmeted Guineafowl (Numidia meleagris) T
I5 ☐	Turkey (Meleagris gallopavo) GC
M4 ☐	Spotted Crake (Porzana porzana)
M5 ☐	Little Crake (Porzana parva)
A ☐	Baillon's Crake (Porzana pusilla)
A ☐	Corncrake (Crex crex)
RB5 ☐	Moorhen (Gallinula chloropus) G,T,GC,F,L
A ☐	Allen's Gallinule (Porphyrula alleni)
A ☐	American Purple Gallinule (Porphyrula martinica)
W2/RB5 ☐	Coot (Fulica atra) T,GC,F
A ☐	Crested Coot (Fulica cristata)
A ☐	Little Bustard (Otis tetrax)
**RB3 ☐	Houbara Bustard (Chlamydotis undulata fuertaventurae) F,L,I
ecE ☐	Canary Islands Black Oystercatcher (Haematopus meadewaldoi)
W/M5 ☐	Oystercatcher (Haematopus ostralegus)
M4 ☐	Black-winged Stilt (Himantopus himantopus)
M5 ☐	Avocet (Recurvirostra avosetta)
**RB4 ☐	Stone-curlew (Burhinus oedicnemus distinctus) H,P,T,GC
**RB2 ☐	(Burhinus oedicnemus insularum) F,L,I
A ☐	Egyptian Plover (Pluvialis aegyptius)
**RB3 ☐	Cream-coloured Courser (Cursorius cursor bannermani) GC?,F,L
M4 ☐	Collared Pratincole (Glareola pratincola)
RB4 ☐	Little Ringed Plover (Charadrius dubius) T,GC,F,L
W/M2 ☐	Ringed Plover (Charadrius hiaticula)
RB2 ☐	Kentish Plover (Charadrius alexandrinus) T,GC,F,L,I
W/M5 ☐	Dotterel (Charadrius morinellus)
A ☐	American Golden Plover (Pluvialis dominica)
W/M5 ☐	Golden Plover (Pluvialis apricaria)
W/M2 ☐	Grey Plover (Pluvialis squatarola)
A ☐	White-tailed Plover (Chettusia leucura)
A ☐	Sociable Plover (Chettusia gregaria)
W/M3 ☐	Lapwing (Vanellus vanellus)
M/W4 ☐	Knot (Calidris canutus)
W/M2 ☐	Sanderling (Calidris alba)
A ☐	Western Sandpiper (Calidris mauri)
M/W4 ☐	Little Stint (Calidris minuta)
M5 ☐	Temminck's Stint (Calidris temminckii)
A ☐	Least Sandpiper (Calidris minutilla)
A ☐	White-rumped Sandpiper (Calidris fuscicollis)
A ☐	Baird's Sandpiper (Calidris bairdii)

A ☐	Pectoral Sandpiper (Calidris melanotos)	
M4 ☐	Curlew Sandpiper (Caladris ferruginea)	
A ☐	Purple Sandpiper (Calidris maritima)	
W/M2 ☐	Dunlin (Calidris alpina)	
A ☐	Buff-breasted Sandpiper (Tryngites subruficollis)	
M4 ☐	Ruff (Philomachus pugnax)	
W/M5 ☐	Jack Snipe (Lymnocryptes minimus)	
W/M3 ☐	Snipe (Gallinago gallinago)	
A ☐	Great Snipe (Gallinago media)	
A ☐	Long-billed Dowitcher (Limnodromus scolopaceus)	
RB4 ☐	Woodcock (Scolopax rusticola) H?,P,G,T	
M/W4 ☐	Black-tailed Godwit (Limosa limosa)	
M/W3 ☐	Bar-tailed Godwit (Limosa lapponica)	
W/M2 ☐	Whimbrel (Numenius phaeopus)	
A ☐	Slender-billed Curlew (Numenius tenuirostris)	
W/M4 ☐	Curlew (Numenius arquata)	
M/W4 ☐	Spotted Redshank (Tringa erythropus)	
M/W3 ☐	Redshank (Tringa totanus)	
A ☐	Marsh Sandpiper (Tringa stagnatilis)	
W/M3 ☐	Greenshank (Tringa nebularia)	
A ☐	Greater Yellowlegs (Tringa melanoleuca)	
A ☐	Lesser Yellowlegs (Tringa flaviceps)	
W4/M3 ☐	Green Sandpiper (Tringa ochropus)	
M3 ☐	Wood Sandpiper (Tringa glareola)	
A ☐	Terek Sandpiper (Xenus cinereus)	
W/M2 ☐	Common Sandpiper (Actites hypoleucos)	
A ☐	Spotted Sandpiper (Actites macularia)	
W/M2 ☐	Turnstone (Arenaria interpres)	
A ☐	Grey Phalarope (Phalaropus fulicarius)	
A(s) ☐	Snowy Sheathbill (Clionis alba)	
M5 ☐	Pomarine Skua (Stercorarius pomarinus)	
M4 ☐	Arctic Skua (Stercorarius parasiticus)	
A ☐	Long-tailed Skua (Stercorarius longicaudus)	
W4 ☐	Great Skua (Stercorarius skua)	
A ☐	Great Black-headed Gull (Larus ichthyaetus)	
A ☐	Mediterranean Gull (Larus melanocephalus)	
A ☐	Laughing Gull (Larus atricilla)	
A ☐	Little Gull (Larus minutus)	
A ☐	Sabine's Gull (Larus sabini)	
A ☐	Bonaparte's Gull (Larus philadelphia)	
W2 ☐	Black-headed Gull (Larus ridibundus)	
A ☐	Slender-billed Gull (Larus genei)	
W5 ☐	Audouin's Gull (Larus audouinii)	
A ☐	Ring-billed Gull (Larus delawarensis)	
A ☐	Common Gull (Larus canus)	
W2 ☐	Lesser Black-backed Gull (Larus fuscus)	
W4 ☐	Herring Gull (Larus argentatus)	
*RB1 ☐	Yellow-legged Gull (Larus cachinnans atlantis)	
W5 ☐	Greater Black-backed Gull (Larus marinus)	
A ☐	Glaucous-winged Gull (Larus glaucescens)	

W4 ☐	Kittiwake (Rissa tridactyla)
M(E)5 ☐	Gull-billed Tern (Gelochelidon nilotica)
A ☐	Caspian Tern (Sterna caspia)
A ☐	Royal Tern (Sterna maxima)
A ☐	Lesser Crested Tern (Sterna bengalensis)
W/M2 ☐	Sandwich Tern (Sterna sandvicensis)
A(b) ☐	Roseate Tern (Sterna dougallii)
M3/SB5 ☐	Common Tern (Sterna hirundo)
M4 ☐	Arctic Tern (Sterna paradisaea)
A ☐	White-cheeked Tern (Sterna repressa)
A ☐	Sooty Tern (Sterna fuscata)
M5 ☐	Little Tern (Sterna albifrons)
M(E)5 ☐	Whiskered Tern (Chlidonias hybridus)
M5 ☐	Black Tern (Chlidonias niger)
A ☐	White-winged Black Tern (Chlidonias leucopterus)
A ☐	Guillemot (Uria aalge)
A ☐	Razorbill (Alca torda)
A ☐	Little Auk (Alle alle)
A ☐	Puffin (Fratercula arctica)
RB3 ☐	Black-bellied Sandgrouse (Pterocles orientalis) F
RB2 ☐	Rock Dove (Columbia livia)
A ☐	Woodpigeon (Columba palumbus)
ecRB4 ☐	Bolle's Pigeon (Columba bollii) H,P,G,T
ecRB5 ☐	Laurel Pigeon (Columba junoniae) P,G,T
I5 ☐	Barbary Dove (Streptopelia risoria)
I?5 ☐	Collared Dove (Streptopelia decaocta) GC,T,L,F
SB2 ☐	Turtle Dove (Streptopelia turtur) all islands
IRB5 ☐	Ring-necked Parakeet (Psitacula krameri) T,GC,F
I5 ☐	Monk Parakeet (Myiopsitta monachus) T,GC,F,L?
A ☐	Great Spotted Cuckoo (Clamator glandarius)
M(E)3 ☐	Cuckoo (Cuculus canorus)
RB3 ☐	Barn Owl (Tyto alba alba) H,P,G,T,GC
**RB4 ☐	(Tyto alba gracilirostris) F,L,I
M5 ☐	Scops Owl (Otus scops)
A ☐	Eagle Owl (Bubo bubo)
A ☐	Tawny Owl (Strix aluco)
**RB3 ☐	Long-eared Owl (Asio otus canariensis) H,P,G,T,GC
W4 ☐	Short-eared Owl (Asio flammeus)
A ☐	African Marsh Owl (Asio capensis)
A ☐	Nightjar (Caprimulgus europaeus)
A ☐	Red-necked Nightjar (Caprimulgus ruficollis)
emRB ☐	Plain Swift (Apus unicolor) all islands
M3,B5 ☐	Common Swift (Apus apus), breeds GC,T?
*RB2 ☐	Pallid Swift (Apus pallidus brehmorum) H?,P,G?,T,GC,F,L.
M4 ☐	Alpine Swift (Apus meelba)
A ☐	White-rumped Swift (Apus caffer)
A ☐	Little Swift (Apus affinis)
A ☐	Kingfisher (Alcedo atthis)
A ☐	Blue-cheeked Bee-eater (Merops superciliosus)
M3 ☐	Bee-eater (Merops apiaster)

M5 ☐	Roller (Coracias garrulus)
RB2 ☐	Hoopoe (Upupa epops)
M(E)4 ☐	Wryneck (Jynx torquilla)
**RB3 ☐	Great Spotted Woodpecker (Dendrocopos major canariensis) T
**RB3 ☐	(Dendrocopos major thanneri) GC
A ☐	Bar-tailed Desert Lark (Ammomanes cincturus)
A ☐	Dupont's Lark (Chersophilus duponti)
A ☐	Calandra Lark (Melanocorhypha calandra)
M4 ☐	Short-toed Lark (Calandrella brachydactyla)
RB5 ☐	Lesser Short-toed Lark (Calandrella rufescens rufescens) T,GC?
**RB1 ☐	(Calandrella rufescens polatzeki) T,GC,F,L,I
A ☐	Crested Lark (Galerida cristata)
W3 ☐	Skylark (Alauda arvensis)
M3 ☐	Sand Martin (Riparia riparia)
A ☐	Crag Martin (Ptyonoprogue rupestris)
M2 ☐	Swallow (Hirundo rustica)
M4 ☐	Red-rumped Swallow (Hirundo daurica)
A ☐	Cliff/Cave Swallow (Hirundo pyrrhoneta/fulva)
M2 ☐	House Martin (Delichon urbica)
A ☐	Richard's Pipit (Anthus novaeseelandiae)
M(E)4 ☐	Tawny Pipit (Anthus campestris)
em**RB1 ☐	Berthelot's Pipit (Anthus berthelotii berthelotii) all islands
M(E)2 ☐	Tree Pipit (Anthus trivialis)
W3 ☐	Meadow Pipit (Anthus pratensis)
M5/W5 ☐	Red-throated Pipit (Anthus cervinus)
A ☐	Water Pipit (Anthus spinoletta)
A ☐	Rock Pipit (Anthus petrosus)
M3 ☐	Yellow Wagtail (Motacilla flava) various races
**RB3 ☐	Grey Wagtail (Motacilla cinerea canariensis) P,G,T,GC
W(E)3 ☐	(Motacilla cinerea cinerea)
W3 ☐	White Wagtail (Motacilla alba)
A ☐	Rufous Bushchat (Cercotrichas galactotes)
RB4/W(E)4 ☐	Robin (Erithacus rubecula rubecula) H,P,G/F,L
**RB4 ☐	(Erithacus rubecula superbus) T,GC
A ☐	Thrush Nightingale (Luscinia luscinia)
M(E)4 ☐	Nightingale (Luscinia megarhynchos)
M(E)5 ☐	Bluethroat (Luscinia svecica)
W4 ☐	Black Redstart (Phoenicurus ochruros)
M(E)3 ☐	Redstart (Phoenicurus phoenicurus)
M(E)3 ☐	Whinchat (Saxicola rubetra)
ecRB3 ☐	Canary Islands Chat (Saxicola dacotiae) F
W5 ☐	Stonechat (Saxicola torquata)
A ☐	Isabelline Wheatear (Oenanthe isabellina)
M2 ☐	Wheatear (Oenanthe oenanthe)
A ☐	Black-eared Wheatear (Oenanthe hispanica)
A ☐	Desert Wheatear (Oenanthe deserti)
A ☐	Hooded Wheatear (Oenanthe monacha)
A ☐	Rock Thrush (Monticola saxatilis)
A ☐	Blue Rock Thrush (Monticola solitarius)
A ☐	Ring Ouzel (Turdus torquatus)

*RB3 ☐	Blackbird (Turdus merula cabrerae) H,P,G,T,GC
A ☐	Fieldfare (Turdus pilaris)
W(E)4 ☐	Song Thrush (Turdus philomelos)
W(E)4 ☐	Redwing (Turdus iliacus)
A ☐	Fan-tailed Warbler (Cisticola juncidis)
M5 ☐	Grasshopper Warbler (Locustella naevia)
A ☐	Aquatic Warbler (Acrocephalus paludicola)
M(E)4 ☐	Sedge Wabler (Acrocephalus schoenobaenus)
M(E)4 ☐	Reed Warbler (Acrocephalus scirpaceus)
A ☐	Great Reed Warbler (Acrocephalus arundinaceus)
M(E)5 ☐	Olivaceous Warbler (Hippolais pallida)
M(E)4 ☐	Melodious Warbler (Hippolais polyglotta)
A ☐	Tristram's Warbler (Sylvia deserticola)
**RB2 ☐	Spectacled Warbler (Sylvia conspicillata orbitalis) all islands
M(E)4 ☐	Subalpine Warbler (Sylvia cantillans)
A ☐	Menetries Warbler (Sylvia mystacea)
RB2 ☐	Sardinian Warbler (Sylvia melanocephala) all islands
A ☐	Desert Warbler (Sylvia nana)
A ☐	Orphean Warbler (Sylvia hortensis)
A ☐	Lesser Whitethroat (Sylvia curruca)
M(E)4 ☐	Whitethroat (Sylvia communis)
M(E)3 ☐	Garden Warbler (Sylvia borin)
RB3 ☐	Blackcap (Sylvia atricapilla heineken) H,P,G,T,GC
M(E)2/W4 ☐	(Sylvia atricapilla atricapilla)
A ☐	Yellow-browed Warbler (Phylloscopus inornatus)
M(E)4 ☐	Bonelli's Warbler (Phylloscopus bonelli)
M(E)3 ☐	Wood Warbler (Phylloscopus sibilatrix)
**RB1 ☐	Chiffchaff (Phylloscopus collybita canariensis) H,P,G,T,GC
**E ☐	(Phylloscopus collybita exsul)
W3/M(E)2 ☐	(Phylloscopus collybita collybita)
W5/M(E)3 ☐	(Phylloscopus collybita brehmii)
A ☐	(Phylloscopus collybita abietinus)
M2 ☐	Willow Warbler (Phylloscopus trochilus)
ecRB3 ☐	Tenerife Goldcrest (Regulus teneriffae) H,P,G,T
M(E)2 ☐	Spotted Flycatcher (Muscicapa striata)
A ☐	Red-breasted Flycatcher (Ficedula parva)
M(E)2 ☐	Pied Flycatcher (Ficedula hypoleuca)
**RB3 ☐	Blue Tit (Parus caeruleus ombriosus) H
**RB3 ☐	(Parus caeruleus palmensis) P
**RB3 ☐	(Parus caeruleus teneriffae) G,T,GC
**RB4 ☐	(Parus caeruleus degener) F,L
M4 ☐	Golden Oriole (Oriolus oriolus)
A ☐	Isabelline Shrike (Lanius isabellinus)
A ☐	Red-backed Shrike (Lanius collurio)
**RB2 ☐	Great Grey Shrike (Lanius excubitor koenigi) T,GC,F,L,I
M(E)3 ☐	Woodchat Shrike (Lanius senator)
A? ☐	Nutcracker (Nucifraga caryocatactes)
RB3 ☐	Chough (Pyrrhocorax pyrrhocorax) P
A ☐	Jackdaw (Corvus monedula)
RB3 ☐	Raven (Corvus corax tingitanus) all islands

W4/RB5 ☐	Starling (Sturnus vulgaris) L,F/T,GC
A ☐	Spotless Starling (Sturnus unicolor)
A ☐	Rose-coloured Starling (Sturnus roseus)
RB1 ☐	Spanish Sparrow (Passer hispaniolensis) all islands
RB5 ☐	Tree Sparrow (Passer montanus) GC
*RB4 ☐	Rock Sparrow (Petronia petronia madeirensis) H,P,G,T,GC
A ☐	Snow Finch (Montifringilla nivalis)
A ☐	Red-billed Quelea (Quelea quelea)
I5 ☐	Common Waxbill (Estrilda astrild) GC,T?
**RB3 ☐	Chaffinch (Fringilla coelebs ombriosa) H
**RB3 ☐	(Fringilla coelebs palmae) P
**RB3 ☐	(Fringilla coelebs tintillon) G,T,GC
A ☐	(Fringilla coelebs coelebs)
ec**RB4 ☐	Blue Chaffinch (Fringilla teydea teydea) T
ec**RB5 ☐	(Fringilla teydea polatzeki) GC
A ☐	Brambling (Fringilla montifringilla)
RB5 ☐	Serin (Serinus serinus) T,GC
emRB2 ☐	Canary (Serinus canaria) H,P,G,T,GC,L,F?
RB3 ☐	Greenfinch (Carduelis chloris) T,GC,F(5),L(5)
RB4 ☐	Goldfinch (Carduelis carduelis) H?,P,G,T,GC,F,L
W5 ☐	Siskin (Carduelis spinus)
**RB1 ☐	Linnet (Acanthis cannabina meadewaldoi) H,P,G,T,GC
**RB1 ☐	(Acanthis canabina harterti) F,L,I
**RB4/RB2 ☐	Trumpeter Finch (Budanetes githagineus amantum) H?,G,T,GC/F,L,I
A ☐	Crossbill (Loxia curvirostris)
A ☐	Waterthrush sp. (Seiurus sp.)
A ☐	Snow Bunting (Plectrophenax nivalis)
A ☐	Cirl bunting (Emberiza cirlus)
A ☐	Rock Bunting (Emberiza cia)
A ☐	House Bunting (Emberiza striolata)
M(E)4 ☐	Ortolan Bunting (Emberiza hortulana)
A ☐	Cretzschmar's Bunting (Emberiza caesia)
A ☐	Little Bunting (Emberiza pusilla)
RB3/RB5 ☐	Corn Bunting (Miliaria calandra) H,P,G,T,GC/F,L

In the following lists * denotes endemic species

MAMMALS

Algerian Hedgehog (Atelerix algirus) F,L,GC,T
*Canary Islands White-toothed Shrew (Crocidura canariensis) F,L
Common White-toothed Shrew (Crocidura osorio) GC,T?
Pigmy White-toothed Shrew (Suncus etruscus) T
Grey Long-eared Bat (Plecotus austriacus)
Barbastelle (Barbastella barbastellus)
Savi's Pipistrelle (Pipistrellus savii)
Kuhl's Pipistrelle (Pipistrellus kuhli)
*Atlantic Islands Pipistrelle (Pipistrellus madeirensis)
(Nyctalus reisleri)
Free-tailed Bat (Tadarida teniotis)
Rabbit (Oryctolagus cuniculus) all islands
Barbary Ground Squirrel (Atlantoxerus getulus) F (introduced)
Black Rat (Rattus rattus) all islands
Brown Rat (Rattus norvegicus) all islands
House Mouse (Mus musculus) all islands
Mouflon (Ovis musimon) T (introduced)
Barbary Sheep (Ammotrapus lervia) GC
Cat (Felis catus) feral on all islands

WHALES AND DOLPHINS
(Nomenclature follows Watson, 1985)

Fin Whale (Balaenoptera physalus)
Minke (Piked) Whale (Balaenoptera acutorostrata)
Bryde's (Tropical) Whale (Balaenoptera edeni)
Cuvier's (Goosebeak) Whale (Ziphius cavirostris)
Northern Bottlenose Whale (Hyperoodon ampullatus)
Gervais's (Gulf Stream) Beaked Whale (Mesoplodon europaeus)
Dense Beaked Whale (Mesoplodon densirostris)
Roughtooth Dolphin (Steno bredanensis)
Shortfin Pilot Whale (Globiocephala macrorhynchus)
Killer Whale (Orcinus orca)
Fraser's (Shortsnout) Dolphin (Lagenodelphis hosei)
Risso's (Grey) Dolphin (Grampus griseus)
Striped Dolphin (Stenella coeruleoabla)
Bridled Dolphin (Stenella attenuata frontalis)
Common Dolphin (Delphinus delphinus)
Bottle-nosed Dolphin (Tursiops truncatus)
Sperm Whale (Physeter macrocephalus)
Pygmy Sperm Whale (Kogia breviceps)
Long-finned Pilot Whale (Globiocephala melaena)

DRAGONFLIES

After Ashmole and Ashmole (1989), Askew (1988)

(Ischnura saharensis)
Emperor dragonfly (Anax imperator)
Lesser emperor dragonfly (Anax parthenope)
Vagrant emperor (Hemianax ephippiger)
(Orthetrum chrysostigma)
Scarlet darter (Crocothemis erythraea)
Red-veined darter (Sympetrum fonscolombei)
(Sympetrum nigrifemur)
(Trithemis arteriosa)
(Zygonyx torridus)

CANARY ISLANDS REPTILES AND AMPHIBIANS
Species and sub-species occurrence in the islands

	REPTILES	Hierro	Gomera	La Palma	Tenerife	Gran Canaria	Fuerteventura	Lanzarote	Other Islands and Rocks	Note
1	Canary Lizard (Firebrand Lizard)									
	Gallotia galloti	CES	CES	CES	CES					
	G g caesaris	IESS (including Roque de Salmor)								
	G g gomerae		IESS							
	G g palmae			IESS						
	G g galloti				IESS (S, SE & Central)					
	G g eisentrauti				IESS (N & NE, Roques de Dentro)					
	G g insulanagae				IESS (Roque de Fuera de Anaga)					
2	Hierro Giant Lizard *Gallotia simonyi*	CES,IES								
3	Gran Canaria Giant Lizard *Gallotia stehlini*					CES, IES				
4	Haria Lizard *Galiotia atlantica*					CES	CES	CES	CES	
	G a delibesi					IESS (in SE)				
	G a atlantica						CESS	CESS	CESS (on Graciosa, Montaña Clara, Lobos)	
	G a laurae							IESS (in NE)	IESS Alegranza	
	G a ibagnezi									
5	Canary Skink (Golden Skink) *Chalcides viridanus*	CES	CES, IESS		CES				+3 populations on small islets	1
	C v viridanu									
	C v caerulopunctatus									

#	Species	Hierro	Gomera	La Palma	Tenerife	Gran Canaria	Fuerteventura	Lanzarote	Other Islands and Rocks	Note
6	Gran Canary Skink — *Chalcides sexlineatus* / *C s sexlineatus* / *C s bistriatus*					CES, IES IESS (South) IESS (North)			+ 2 populations on small islets	1
7	Eastern Canary Skink — *Chalcides polylepis occidentalis*						IESS			2
8	Canary Gecko — *Tarentola delalandii*	MES,CESS (only Roque de Salmar)		MES CESS	MES CESS					3
9	Boettger's Gecko — *Tarentola boettgeri* / *T b boettger* / *T b hierrensis*	MES IESS				MES IESS				
10	Eastern Canary Gecko — *Tarentola angustimentalis*						CES	CES	CES (on Graciosa, Roque de Este & Roque de Oeste)	
11	Gomera Gecko — *Tarentola gomerensis*		CES,IES							
12	Turkish Gecko — *Hemidactylus turcicus*				X	X				4
13	Leatherback Turtle — *Dermochelys coriacea*	Widespread but sporadic, the largest turtle in the waters around the Canaries								
14	Loggerhead Turtle — *Caretta caretta*	Widespread and frequent, much the commonest species around the Canaries								
15	Hawksbill Turtle — *Eretmochelys imbricata*	Occasional. Tropical species, only rarely occurs in northern Europe								
16	Green Turtle — *Chelonia mydas*	Only very occasional. Warm water species. Very rarely seen in European Waters								
AMPHIBIANS										
1	Marsh Frog — *Rana perezi*	X	X	X	X	X	X?	X?		4
2	Stripeless Tree Frog — *Hyla meridionalis*	X	X	X	X	X	X	X		4

NOTES

1. Introduced on Madeira
2. Another sub-species occurs in Morocco
3. Only one other sub-species and this is confined to the Salvage islands
4. Introduced species

KEY

CES	Canary Islands Endemic Species	IESS	Single Island Endemic Sub-species
MESS	Macaronesian Endemic Sub-species	CESS	Canary Islands Endemic Sub-species
MES	Macaronesian Endemic Species	IES	Island Endemic Species
		X	Species occurs on island

SE South-east   N North
NE North-east   S South

CANARY ISLANDS BUTTERFLIES
Species and sub-species occurrence in the islands (see Key on page 109)

#	Common name	Scientific name	Hierro	Gomera	La Palma	Tenerife	Gran Canaria	Fuerteventura	Lanzarote	Note
1	Large White	Pieris brassicae cheiranthi	X	CESS	CESS	CESS	X	X	X	1
2	Small White	Artogeia rapae	X	X	X	X	X	X	X	
3	Bath White	Pontia daplidice	X	X	X	X	X	X	X	
4	Green-striped White	Euchloe belemia eversi				IESS				
		Euchloe belemia hesperidum					CESS	CESS		
5	Greenish Black-tip	Elphinstonia charlonia charlonia				X but none recently		X	X	
6	African Migrant	Catopsilia florella		X		X	X			
7	Clouded Yellow	Colias crocea	X	X	X	X	X	X	X	
8	Cleopatra ("Canary Island Brimstone")	Gonopteryx cleopatra cleobule		CESS		CESS				1
		Gonopteryx cleopatra palmae			IESS					
9	Small Copper	Lycaena phlaeas phlaeas	X	X	X	X	X	X	X	
10	Long-tailed Blue	Lampides boeticus	X	X	X	X	X	X	X	
11	Canary Blue	Cyclyrius webbianus	CES	CES	CES	CES	CES			
12	African Grass Blue	Zizeeria knysna knysna	X	X	X	X	X	X	X	
13	(Southern) Brown Argus	Aricia (agestis) cramera		X	X	X	X			
14	Common Blue	Polyommatus icarus				X	X	X	X	2
15	Diadem	Hypolimnas misippus		X		X				
16	Red Admiral	Vanessa atalanta	X	X	X	X	X	X	X	
17	Indian Red Admiral	Vanessa indica vulcania	MESS	MESS	MESS	MESS	MESS	MESS		
18	Painted Lady	Cynthia cardui	X	X	X	X	X	X	X	
19	American Painted Lady	Cynthia virginiensis	(X)	(X)	(X)	(X)	(X)			3
20	Cardinal	Pandoriana pandora (seitzi?)		CESS	CESS	CESS				4
21	Queen of Spain Fritillary	Issoria lathonia		X	X	X	X			
22	Canary Grayling	Pseudotergumia wyssii wyssii			(CES??)	CES	CES			
		Pseudotergumia wyssii bacchus	CES	CES						
23	Meadow Brown	Maniola jurtina hispulla	X	X	X	X	X			
24	Canary Speckled Wood	Parage xiphioides		CES	CES	CES	CES			
25	Monarch	Danaus plexippus	X	X	X	X	X			
26	Plain Tiger	Danaus chrysippus		X	X	X	X	X	X	
27	Lulworth Skipper	Thymelicus acteon christi		CESS	CESS	CESS	CESS			

NOTES

1. Sometimes this race considered as separate species; if so then CES
2. Sometimes this race considered as separate species but if so not endemic
3. Existence of recent records uncertain (post 1970)?
4. Endemic sub-species status disputed

Adapted from information in Ashmole M & Ashmole P 1989, Higgins H G & Hargreaves B 1983 and Fernández-Rubio F 1991.